TESLA

TESLA

INVENTOR OF THE MODERN

Richard Munson

W. W. NORTON & COMPANY
Independent Publishers Since 1923

For information about permission to reproduce selections from this book, write to
Permissions, W. W. Norton & Company, Inc., 500 Fifth Avenue, New York, NY 10110

For information about special discounts for bulk purchases, please contact
W. W. Norton Special Sales at specialsales@wwnorton.com or 800-233-4830

Manufacturing by LSC Communications, Harrisonburg
Book design by Daniel Lagin
Production manager: Julia Druskin

Library of Congress Cataloging-in-Publication Data

Names: Munson, Richard, author.
Title: Tesla : inventor of the modern / Richard Munson.
Description: First edition. | New York : W. W. Norton & Company, [2018] | Includes
bibliographical references and index.
Identifiers: LCCN 2017055596 | ISBN 9780393635447 (hardcover)
Subjects: LCSH: Tesla, Nikola, 1856–1943. | Electrical engineers—United States—Biography. |
Inventors—United States—Biography. | Electrical engineering—History.
Classification: LCC TK140.T4 M86 2018 | DDC 621.3092 [B]—dc23
LC record available at https://lccn.loc.gov/2017055596

ISBN 978-0-393-35804-9 pbk.

W. W. Norton & Company, Inc., 500 Fifth Avenue, New York, N.Y. 10110
www.wwnorton.com

W. W. Norton & Company Ltd., 15 Carlisle Street, London W1D 3BS

2 3 4 5 6 7 8 9 0

To Kathryn

We wind a simple ring or iron with coils; we establish the connections to the generator, and with wonder and delight we note the effects of strange forces which we bring into play, which allow us to transform, to transmit, and direct energy at will.

NIKOLA TESLA, 1892

CONTENTS

TESLA

INTRODUCTION
EVERYWHERE IS ENERGY

Nikola Tesla is known as an eccentric genius, even a visionary. The question is, did his eccentricities eclipse his genius? Does it matter? Larry Page, a founder of Google, praises him as "a hero." Elon Musk's car and company bear his name.

Entrepreneurs like Page and Musk—multibillionaires and brilliant discoverers—point to Nikola Tesla as the one who launched our modern era of electricity, radio, and robots. They consider him an inventor's inventor, something of a folk legend.

Tesla's electric motors run our appliances and factories, so why is Thomas Edison the more famous of the two? Likewise, Guglielmo Marconi gets popular credit for inventing the radio, yet the U.S. Supreme Court ruled that Tesla's patents laid the foundation for transmitting wireless signals over long distances.

Who was this farsighted, if underappreciated, mastermind?

Perhaps the best introduction occurred in 1891, on a spring night in New York City, when hundreds crammed into an auditorium on the Columbia College campus to witness a key battle in the "War of the Currents." Serbian-born Nikola Tesla had declared that his method of harnessing electricity could outperform and outdistance Thomas Edison's system. Once an employee and now a rival of Edison, Tesla also claimed he could transmit sound with electrical charges. The scientists and engineers came to see Tesla—inventor and scientific magician—wield his wondrous power.

The thirty-five-year-old Tesla was invited to the stage by the two Columbia School of Mines professors—the university had only started an electrical engineering program two years earlier. They'd arranged for Tesla to install his innovative high-frequency alternator offstage, in a nearby building called the "cowshed." The plan was for Tesla to unveil his latest discoveries with great showmanship in front of a huge audience.

Tesla expected controversy from the crowd, which included several Edison supporters. Edison's direct current, or DC, and incandescent bulbs had become the standard and were backed by J. P. Morgan and other powerful bankers. Twelve years had passed since Edison introduced his lamp with a glowing carbon filament within a vacuum and nine years since he opened the first central power station on Pearl Street in New York City. Yet DC's one-directional charge could be distributed for only a mile or so, and Edison's lamps were notoriously inefficient and prone to burning out. Could Tesla's new ideas overcome Edison's financial and commercial advantages?

Nearly a decade younger than the self-taught Edison, Tesla had an impressive and extensive education. He wrote poetry, recited complete books from memory, and spoke eight languages—English, Serbo-Croatian, Czech, Hungarian, German, French, Italian, and Latin. This polyglot reveled in the excitement of discovery. "I do not think there is any thrill that can go through the human heart like that felt by the inventor as he sees some creation of the brain unfolding to success," he said. "Such emotions make a man forget food, sleep, friends, love, everything."[1]

Tesla had already challenged Edison and his supporters in publications, claiming he could build bigger generators, longer transmission lines, and more reliable lamps. In opposition to Edison's direct current, Tesla championed alternating current, or AC, which periodically reversed the direction of the electrical charge. His breakthrough, the Tesla coil, could reliably deliver high-frequency, high-voltage electricity.

And with that ability, Tesla predicted, companies could transmit power over long distances.

The eager crowd arrived early at the auditorium—located in a gray-slab Greek Revival building between Madison and Park Avenues on 49th Street. Having heard about Tesla's man-made lightning, they wanted to see if someone subjected to ten thousand volts, delivered via alternating current, would burst into flames or shoot sparks from his fingertips. Despite the circus-like atmosphere, such lectures were formal affairs, with the audience of male scientists in black suits; a few accompanying wives accented the scene with an occasional plume or lace jabot.

After a short introduction by one of the professors, Tesla approached the stage haltingly. He was a slender man who dressed with a European formality, and for this demonstration he wore his usual dapper four-button, dark-brown cutaway suit, a white monogrammed silk shirt, gray suede gloves, and a black tie, arranged in the old-fashioned, four-in-hand style. At six feet three inches, he towered over his introducer. Tesla had a thick, neatly trimmed mustache, angular face, and wavy hair severely parted down the middle. Surviving severe bouts of malaria and cholera in his teens had left Tesla with a lifelong fear of germs. Walking onstage, he avoided shaking hands, keeping his own clasped behind him, but he bowed politely to a few colleagues. His deep-set, pale, and glimmering eyes stilled the audience.

When he opened his mouth, his high, almost falsetto voice must have surprised the crowd; he spoke in what he called "pure, nervous English." Tesla began by praising a few of the distinguished scientists, including two he knew felt threatened by his new system. More surprisingly, he admitted that he did not fully understand electricity. "Of all the forms of nature's immeasurable, all-pervading energy, which ever and ever change and move, like a soul animates an innate universe, electricity and magnetism are perhaps the most fascinating."[2]

With a joyful smile, Tesla declared: Electricity—what could be more mysterious and useful?

His long, thin hands trembled with enthusiasm as he drew diagrams and formulas on a large blackboard, proving the superiority of high-frequency alternating current that could be sent hundreds of miles. He predicted how increasing the oscillations within an alternating current produced new forms of energy and enabled wireless communications, predating Marconi's discoveries by a decade.

Like a seasoned actor, the usually shy Tesla sensed the crowd's growing interest in his props—an assortment of tubes and bulbs arrayed on a long wooden table at the front of the stage between two large zinc plates suspended from the ceiling. The inventor transformed himself into a showman. Throwing a switch connected to his motor and alternator out in the "cowshed," Tesla cranked up the current's frequency. An electric arc surged across two poles, creating purplish sparks and loud bursts of crackles. Tesla increased the oscillations, and the tone became smoother and higher pitched, the streamers turning a radiant white— in front of a captivated audience, he created lightning on stage. Spark-generated ozone scented the air, what some likened to chlorine bleach and others to the odor of wet hay. The soaring streamers produced a wind felt by those sitting in the first several rows.

The showstopper was still to come. Tesla waved gas-filled tubes between the electrified sheets of zinc placed on both sides of the stage. Within the electrostatic field created by those plates, which were about fifteen feet apart, the tubes glowed. No wires, no flames, no heat, and yet the gas in the tubes fluoresced. The tubes appeared to one reporter "like a luminous sword held in the hand of an archangel representing justice."[3] Another predicted wireless lighting would "bring fairy-land within our homes."[4]

Even Tesla recognized his growing impact on the audience, several of whom considered electricity an occult force. "It is difficult to appreciate what those strange phenomena meant at the time," he observed later. "When my tubes were first publicly exhibited, they were viewed with amazement impossible to describe."[5]

"Mr. Tesla seemed to act the part of a veritable magician," reported *Electrical Review*. "It seemed to make little difference whether the lamps were lying on the table or whether they were connected by one terminal to one pole of the coil, or whether the lecturer took a lamp in each hand and held one to each pole of the coil.... In each and every case the filaments were brought to incandescence, to the supreme delight of the spectators."[6]

Tesla's wonders promised more than profits. "All around us everything is spinning, everything is moving, everywhere is energy," declared the scientist. "There *must* be some way of availing ourselves of this energy directly." With such an achievement, he said, "Humanity will advance with great strides.... [The] mere contemplation of these magnificent possibilities expands our minds, strengthens our hopes, and fills our hearts with supreme delight."[7]

Within the cheering crowd, there were some hecklers. Michael Pupin led a small band of scientists who raised a ruckus. Pupin was a fellow Serbian immigrant, a competitive and egotistical researcher who believed he had developed a better motor that could run on alternating current, albeit with a design based on Tesla's patents. While "I was lecturing," Tesla complained later, "Mr. Pupin and his friends interrupted ... by whistling, and I had difficulty quieting down the misled audience."[8]

But the vast majority of those assembled proved to be far more interested in technological drama than personal squabbles. Tesla's finale featured tens of thousands of volts of alternating current passing through his own body: sparks indeed flew from his fingertips! Rebutting Edison's claims about the dangers of alternating current, the inventor declared that his AC was controlled electricity. He claimed the high-frequency, low current he generated was no more harmful than vibrations of light and that it remained on the surface of his skin, causing no harm to his body.

The lecture featured more than sparkling showmanship, as Tesla systematically described novel laws of electricity. Scientists had been used to certain ways power moved when currents were of a steady character. However, when those currents rapidly changed direction, when they alternated, new rules applied.

Tesla presented his audience with an astounding list of practical applications. Well before others, he sensed the promise of alternating current. On stage, he'd transmitted electricity wirelessly by charging two plates fifteen feet apart. How much further could energy travel wirelessly? By increasing a current's fluctuations, he suggested his high-frequency alternator would send and receive messages and sounds. He described long-distance electricity transmission, powerful motors, and laborsaving appliances. While not using our modern words, he foresaw—and later developed—radio, robots, and remote control.

Tesla performed for three hours that night. Perfecting the role of entertainer, he concluded by saying that with more time he would have revealed even more revolutionary experiments that he had conducted in his laboratory. He left the audience, and possible investors, applauding loudly . . . and wanting more.

Electrical Review hailed the event as "brilliant" and predicted attendees would "remember that occasion as one of the scientific treats of their lives."[9] Another reporter suggested Tesla had "eclipsed" Edison with a better incandescent bulb and gone beyond all other scientists with improved light-producing vacuum tubes. *Harper's Weekly* declared that Tesla "in one bound" had joined the ranks "of such men as Edison, [Charles] Brush, Elihu Thomson, and Alexander Graham Bell." Both Brush and Thomson were famous for developing electrical generators and arc-lighting systems that brightened city streets. Describing Tesla's rags-to-riches story, the magazine added, "Yet only four or five years ago, after a period of struggle in France, this stripling from the dim mountain border-land of Austro-Hungary landed on our shores, entirely unknown, and poor in everything save genius and training, and courage."[10]

Not everyone was impressed. The English journal *Industries* questioned Tesla's devotion to vision over practicality, a complaint that echoed throughout the scientist's life. "Anyone who had read many of Mr. Tesla's articles," the magazine wrote, "must have difficulty in understanding the frequent vague and idiomatic statements with which they abound."[11]

In fact, what the critic objected to may be a result of Tesla's singular talent. There are many stories of him visualizing his inventions so clearly and completely that his articles may sound "vague and idiomatic" in comparison to the diagrams in his head. Tesla was actually born during a lightning storm and his earliest memories were of bright hallucinations that blurred his sense of reality. Yet displaying one of many paradoxes, he also possessed the rare ability to see and develop complex equipment in his thoughts, frequently needing neither models nor adjustments in order to realize his detailed inventions. His genius may have been his ability to sort through the various bright visions filling his head and devise practical devices or prophetic ideas.

This prolific maverick foresaw cell phones, radar, laser weapons, artificial intelligence, the Internet, fax machines, and vertical-lift airplanes. In his lifetime, Tesla obtained some three hundred patents and provided our modern economy with electric motors, robots, remote controls, and radio. According to the American Institute of Electrical Engineers at the cusp of the twenty-first century, "Were we to seize and eliminate from our industrial world the results of Mr. Tesla's work, the wheels of industry would cease to turn, our electric cars and trains would stop, our towns would be dark, our mills would be dead and idle."[12]

Tesla also circulated quixotic notions that have complicated his legacy. He sketched plans for communicating with intelligent beings on other planets, reading another person's thoughts by attaching television equipment to her retina, and sending electricity wirelessly through the earth so everyone could enjoy power at virtually no cost.

Perhaps it's not surprising, as noted in his *New York Times* obituary, that this distracted discoverer was "anything but a practical man as far as business was concerned."[13] He was bested by robber barons.

Yet among inventors, even some seventy-five years after his death, Tesla's creativity and industriousness continue to inspire. Elon Musk recently donated one million dollars to restore the scientist's Long Island laboratory into a museum. Perhaps the inventor's reputation can be restored as well, and Nikola Tesla will be accorded the fame he deserves.

1.

BORN BETWEEN TODAY
AND TOMORROW

Lika

I s it apocryphal—or appropriate—that Nikola Tesla was born in the midst of a violent thunderstorm exactly at the stroke of midnight, between the ninth and tenth of July 1856?[1] The frightened midwife feared he would be "a child of the storm." His mother, however, declared, "No, of light."[2]

The weirdness of his birth during a midnight gale became part of Tesla family lore, no doubt assuring the young boy he possessed special qualities. His father was a Serbian Orthodox priest, and when his beloved mother, Djuka, repeated the tale, she drew references to her religion's Holy Fire and the symbolic power of lighting candles to overcome the darkness. In fact, light pervaded Tesla's life, filling his mind with both inspirational and bizarre visions. The birth story also placed Nikola Tesla in the present as well as the future, neither today nor tomorrow.

Ambiguity and mutability molded Tesla's personality. He was born a Serb in what today is Croatia. The vast majority of his neighbors attended Roman Catholic churches, while his family practiced Serbian Orthodox Christianity, making Tesla a member of a religious as well as an ethnic minority. While their small village was under the rule of the Habsburg Monarchy, the region faced constant change and would soon be part of the Austro-Hungarian Empire.

Growing up Serbian evoked not only ethnic pride but also a tragic past—another of Tesla's many paradoxes. Serbian national poetry, according to Tesla, was "full of the admiration for the feats of the heroes."[3] Although Balkan history is complicated and much debated, Tesla focused on what he felt was his region's most heartbreaking chapter: "Hardly is there a nation which has met with a sadder fate than the Serbians." Their once splendid empire was "plunged into abject slavery after the fateful battle of 1389 at the Kosovo Polje, against the overwhelming Asian hordes."[4] On that day—considered as significant to the Serbs as the Exodus is to the Jews—some thirty thousand Turks obliterated the Serbian nation, converted its churches into mosques, enslaved its males into the Turkish army, and forbid Serbs from owning property or learning to read, write, or play musical instruments. Nearly five hundred years in the past, that bloody event—the memory of which unified the Serbian people's identity—still prompted Serbs (and Tesla) to believe "Europe can never repay the great debt it owes to the Serbians for checking, by the sacrifice of its own liberty, that barbarian influx."[5] That event also prompted Serbians to embody "in immortal song" the "brave deeds of those who fell in the struggle for liberty." Tesla proudly noted Serbia became "a nation of thinkers and poets."[6]

(The 1389 Battle of Kosovo took place on June 15, which became known as St. Vitus Day. Two other notable events occurred on that date: the Serbian 1876 declaration of war against the Ottoman Empire and the 1914 assassination in Sarajevo of Archduke Franz Ferdinand of Austria, which ignited World War I.)

Tesla's immediate family got its start in Senj, a small seaside town, where senior clerics had assigned his parents—the twenty-eight-year-old Milutin Tesla and the twenty-five-year-old Djuka Mandic—after their 1847 wedding. Tesla's father was the novice priest there, serving some forty households at the stone church perched on a steep cliff overlooking the Adriatic Sea, and he was to represent Serbs before "foreign and Catholic persons." Milutin's mediating role confronted stark historical and cultural differences. The majority Croats, for instance, followed the Roman form of Catholicism and embraced the Pope as their spiritual leader, whereas Serbs worshiped in Greek

Orthodox churches and adopted a Byzantine patriarch. Croats, moreover, utilized the Latin alphabet while Serbs primarily wrote in Cyrillic.

Milutin made a name for himself as a fervent Serbian advocate, writing in several journals about the need to preserve Serbian traditions and advance their political and social independence. When Milutin tried to have Serbian soldiers attend Orthodox services on Sundays, the Austrian military commander refused and ordered the Serbs to participate in Roman Catholic Mass instead. "Nothing is as sacred to me as my church and my forefathers' law and custom," Milutin penned in a letter, "and nothing so precious as liberty, well-being, and advancement of my people and my brothers, and for these two, the church and the people, wherever I am, I'll be ready to lay down my life."[7] He periodically wrote for the *Serbian Daily*, including one piece that boasted: "This Serbian valley is blessed with the greatest wealth in the world— its people are robust and heroic in every respect."[8]

Djuka gave birth in Senj to the first three (Milka, Dane, and Angelina) of their five children, yet Milutin's salary proved trifling and the damp air troubled his health. After numerous attempts over eight years, the zealous preacher obtained a transfer in 1852 to the church of St. Apostles Peter and Paul, a weather-beaten, white structure within the village of Smiljan and the province of Lika, where Milutin served a larger congregation of eighty households. It was in this more populated but still remote Croatian province in 1856 that Djuka delivered Nikola Tesla, the couple's fourth child. Their daughter Marica was also born in Lika three years later.

Tesla's elder brother, Dane, had been his idol and the family's favorite child. Considered "gifted to an extraordinary degree,"[9] Dane was expected to follow his father into the clergy.

Daily life in Smiljan—despite the name meaning "the place of sweet basil"—was filled with drudgery. Tesla's neighbors said of their hard-to-farm homeland: "When God distributed the rocks over the earth He carried them in a sack, and when He was above our land the sack broke."[10] They cleared those plentiful stones from the fields so crops could be tended with hand tools. To obtain illumination, warm their rooms, or cook their meals, villagers had to chop trees down with axes and carry the wood in slings and carts to their houses. Far from any city, they endured remoteness, particularly throughout the long winters.

Tesla's mother tackled that drudgery and isolation by developing several laborsaving devices, including a mechanical eggbeater. Inventing out of necessity, she set a model for her son. With an aesthetic inventiveness, Djuka spun threads, created intricate designs, and fabricated most of the family's clothing and home furnishings. Tesla wrote that even in her sixties, "her fingers were still nimble enough to tie three knots in an eyelash."[11] Later he declared, "I must attribute to my mother's influence whatever inventiveness I possess."

Tesla also credited his mother with his own industriousness. She was, he wrote, "indefatigable and worked regularly from four o'clock in the morning till eleven in the evening. From four to breakfast time, 6 a.m., while others slumbered, I never closed my eyes but watched my mother with intense pleasure as she attended quickly—sometimes running—to her many self-imposed duties. She directed the servants to take care of all domestic animals, milked the cows, performed all sort of labor unassisted, set the table, prepared breakfast for the whole household and only when it was ready to be served did the rest of the family get up. After breakfast everybody followed my mother's inspiring example. All did their work diligently, liked it, and so achieved a measure of contentment. But I was the happiest of them."[12]

Accentuating his parental preference between his unlettered mother and learned father, Tesla admitted the training Milutin provided "must have been helpful," but he stressed his mother's accomplishments: "She was a truly great woman, of rare skill, courage, and fortitude, who had braved the storms of life and passed through many a trying experience."[13] Djuka could neither read nor write, and yet she could recite from memory long passages from the Bible and epic Serbian poems. Tesla argued she would "have achieved great things had she not been so remote from modern life and its multifold opportunities."[14]

Djuka herself was descended from a distinguished family of priests and inventors. Her grandfather, a cleric, received the Medal of Honor from Napoleon himself for helping the French to occupy Croatia. One of her brothers became the Archbishop of Sarajevo and Metropolitan of the Serbian Orthodox Church in Bosnia, while both her father and grandfather devised numerous farm and household tools.

Tesla's father came from a military family and often told tales of tragic Serbian war heroes though he failed to follow the tradition of his ancestors, unlike his own father and brother. The army's discipline did not suit Milutin, so after being reprimanded for not polishing his brass buttons sufficiently, he resigned and enrolled in an Orthodox seminary, where he graduated top of his class in 1845. As his son later remarked, Milutin "had a prodigious memory and frequently recited at length from works in several languages." The family, in fact, joked that "if some of the classics were lost, he would be able to restore them." Considered something of a progressive, Milutin took up several political causes, including compulsory education for children, schools where Serbs could learn their own language, and social equity among ethnic groups.[15]

Tesla found his father to be "a very erudite man, a veritable natural philosopher, poet, and writer."[16] Milutin composed poetry and editorials with a style of writing Tesla later said was "much admired [and with] sentences full of wit and satire." Milutin also spoke multiple languages, was an omnivorous reader, and proved to be an accomplished mathematician. Yet the father was more religious than scientific, and he consistently pressured his son to follow his own example. Tesla did not resemble his father; Milutin was pale and displayed a receding hairline, shiny forehead, high cheekbones, and a full beard. According to Tesla, Milutin also "had the odd habit of talking to himself and would often carry on an animated conversation and indulge in heated argument, changing the tone of his voice."

The father could be stern and disciplinary toward his children but show others a sense of humor. When driving with a friend whose costly fur coat rubbed against the carriage wheel, for example, Milutin quipped, "Pull in your coat, you are ruining my tyre."[17]

Nikola was named after both grandfathers. The Tesla name, in Serbian, has two somewhat related meanings. The first describes a person with protruding teeth, which was true of several family members but not of Nikola; the second references an adze, a cutting tool used in woodworking that looks something like a strong upper jaw.

The young Tesla spent a great deal of time with his two older sisters in a childhood he described as "blissful." He played with pets and rode atop his father's Arabian horse, which had been a gift from a family friend. Not so the farm's gander, which he later described as "a monstrous ugly brute, with a neck of an ostrich, mouth of crocodile, and a pair of cunning eyes radiating intelligence and understanding like the human." This "powerful enemy" once seized the infant and "almost [pulled] out the remnant of my umbilical cord."[18] On another occasion, the gander attacked Tesla as he entered the poultry yard, grabbing "me by the seat of my trousers and shaking me viciously. When I finally managed to free myself and run away, he would flap his huge wings in glee and raise an unholy chatter in which all the geese joined."[19]

Family members recalled funny and revealing stories of young Tesla. One involved two old aunts with wrinkled skin, "one of them having two teeth protruding like the tusks of an elephant." These relatives had the habit of hugging Tesla aggressively, and one regularly buried her face, protruding teeth and all, into his neck. According to Tesla, they were "as affectionate as they were unattractive." While being carried one day in his mother's arms, he was asked by the aunts to choose which of them was prettier. Suggesting thoughtfulness (if not diplomacy), after examining their faces carefully, he declared, "This one here is not as ugly as the other."[20]

When one of the aunts suggested Tesla feared the family cow, he proceeded that afternoon to demonstrate his bravery by climbing a fence and sliding onto the creature's "back for a ride when she made off with me bellowing and threw me." Still, recalled Tesla, "I was none the worse for the experience."[21]

Tesla's home was an "old-fashioned structure located at the foot of a wooded hill called Bogdanic." According to Tesla, "Adjoining it on the other side is a church and behind the same, further up, a graveyard. Our nearest neighbors were two miles away and in the winter, when the snow was six feet deep, if not more, our isolation was complete."[22] (Noting the inventor's continued popularity in his native country, the buildings have been converted into a popular tourist attraction, located outside the still quiet hamlets of Smiljan some ten minutes from a highway exit.)

Isolation remained a theme throughout Tesla's life. The remoteness that may have necessitated his mother's ingenuity also impeded her fame and convinced Tesla he needed to move away in order to pursue his engineering goals. He did not see a steam locomotive or other industrial machines until he was a teenager. Yet Lika's isolation garnered within Tesla a self-reliance, an independence, and throughout his life he sought solitude in order to envision his revolutionary designs.

The natural setting of Lika also prompted Tesla to observe carefully, and such close attention often led to profound insights. A snowball gaining size as it rolled down a hill gave rise to the concept of "magnification of feeble actions," prompting Tesla later to consider using the earth's resonance to amplify the impact of an electrical thrust. The appearance in order of lightning, thunder, and torrents convinced the sensitive young boy of the sequencing of nature's secrets. Sensing the power that could be tapped from cascading water, he also observed that if warm winds from the Adriatic melted the winter's snow rapidly, "we would then witness the terrifying spectacle of a mighty seething river carrying wreckage and tearing down everything movable in its way."[23]

Tesla's greatest youthful joy came from "our magnificent Macak— the finest of all cats in the world." Maybe a lack of companionship prompted such devotion. "We simply lived one for the other," he recalled. "Wherever I went Macak followed primarily owing to our mutual love and then again moved by the desire to protect me.... He fascinated me so completely that I too bit and clawed and purred.... [On rainy days] we went into the house and, selecting a nice, cozy place, abandoned ourselves to each other in affectionate embracement."[24]

Tragedy broke the tranquility when Tesla was seven years old. It seems the family's beloved but high-spirited Arabian stallion bolted and threw the twelve-year-old Dane, who died that night from his injuries.

(Some biographers have proposed that Tesla spooked the horse or even knocked his brother down a flight of stairs, but there's no evidence for such claims and Tesla's reactions suggest otherwise.)

Young Tesla witnessed the terrible scene but offered few details about his brother's accident. What he best remembered was his mother waking him in the middle of that night. "It was a dismal night with rain falling in torrents," he recalled. "My mother came to my room, took me in her arms and whispered almost inaudibly: 'Come and kiss Dane.' I pressed my mouth against the ice-cold lips of my brother knowing only that something dreadful had happened. My mother put me again to bed and lingering a little said with tears streaming: 'God gave me one at midnight and at midnight he took away the other one.'"[25] More than fifty-five years later, Tesla still lamented, "My visual impression of it has lost none of its force."

Having always favored Dane, Milutin and Djuka increasingly idolized the dead boy's talents and consistently projected accomplishments if he had lived. Tesla could not compete with such an ideal. "Anything I did that was creditable merely caused my parents to feel their loss more keenly," he later wrote. "I grew up with little confidence in myself."[26] Tesla lived with the awkward knowledge that any success he earned reminded his parents of the achievements Dane could not claim.

Although Tesla was precocious—probably at least as talented as Dane—he felt inadequate. "The recollection of [Dane's] attainments makes every effort of mine seem dull in comparison," he admitted. The message from his parents, especially his father, was "don't succeed." Yet Tesla disclosed that the loss of his brother—and the ongoing failure to please his parents—pushed him to discipline his thoughts and actions.

No doubt Tesla knew he was "far from being considered a stupid boy." He recalled playing with friends in the street and encountering a wealthy alderman who gave a silver piece to each of the boys. When the gentleman approached Tesla, however, he stopped and demanded, "Look in my eyes." Tesla did as he was told, held out his hand for a coin, but disappointedly heard: "No, you can get nothing from me, you are too smart."[27]

Losing their first-born son particularly affected Milutin. Although the father continued preaching, he wrote fewer and fewer articles and poems. Trying to escape the sad memories of Lika, Milutin moved his family to Gospic, the region's administrative center. Tesla had loved the freedom of the farm's open spaces, and he disliked the larger town's pace and pressure. He particularly missed his pigeons, chickens, and sheep. "In our new house I was but a prisoner," he wrote, "watching the strange people I saw through the window blinds. My bashfulness was such that I would rather have faced a roaring lion than one of the city dudes who strolled about."[28]

Two mishaps marked Tesla's introduction to his new home. In the first, after one of his father's sermons, Tesla finished ringing the church bells in the belfry and rushed downstairs, accidentally tripping on the enormous train of a dress worn by the town's wealthiest lady; the back of the gown "tore off with a ripping noise which sounded like a salvo of musketry fired by raw recruits." The "good and pompous woman" expressed great displeasure at the destruction of her elegant clothing, and Tesla's father, "livid with rage," gave him a "gentle slap on the cheek, the only corporal punishment he ever administered to me." Deeply embarrassed, Tesla felt further ostracized from both his father and the community.[29]

The second mishap, however, provided some relief. The town's fire department had purchased a new truck and pump, whose operation required the efforts of sixteen men. On the day of its ceremonial inauguration, a crowd assembled to hear speeches, yet the pump didn't work. While the firemen and the school's teachers debated furiously about how to fix the problem, Tesla waded into the river and "instinctively felt for the suction hose in the water and found that it had collapsed." When he loosened the knot, water rushed forth, the crowd cheered, and the firemen carried Tesla aloft on their shoulders and declared him to be "the hero of the day."[30]

Milutin used the new location to launch an aggressive mind-training campaign, drilling his surviving son with "all sorts of exercise, such as guessing one another's thoughts, discovering the defects of some form or expressions, repeating long sentences or performing mental calculations." The restless young Tesla probably disliked the discipline, but later in life he reflected, "These daily lessons were intended to strengthen memory and reason and especially to develop the critical sense."[31]

Milutin intended his son to be a cleric, in part to protect his only male heir from the dangers of the military and the rigors of intensive engineering study. Although the religious prospect "constantly oppressed" Tesla, who "longed to be an engineer," his father proved to be inflexible. Milutin seemed hell bent on not letting Nikola be Nikola.

At the age of ten, Tesla entered Gospic's Normal School, which added to his father's regimentation. He excelled at mathematics and sciences and proved capable of performing calculations in his mind, prompting allegations of cheating from his teachers. The school introduced Tesla to mechanical models, with a water turbine attracting his special interest. Upon seeing an engraving of Niagara Falls in the classroom, he claimed to have informed an uncle that he would one day travel to the United States and construct "a big wheel run by the Falls." The young student devised "all kinds of contrivances and contraptions" but was particularly pleased with his crossbows and slings, which his teachers "judged to be the best in his class."[32]

Tesla tinkered joyously—and often got into trouble. He relished taking apart his grandfather's clocks but usually reassembled them incorrectly, provoking his grandfather to bring "my work to a sudden halt in a manner that was not too delicate." He developed a fairly creative pop gun—comprised of a hollow tube, a piston, and two plugs of hemp—and showed substantial shooting skill, but, as he later admitted, "My activities interfered with the window panes in our house, and soon met with the most painful discouragement." Young Tesla also acted out the role of Serbian heroes. Sword in hand, he mowed down enemies in the form of corn stalks, until his mother realized these battles were ruining her plants. She responded with several spankings, "not of the formal kind, but a genuine article."[33]

As early as ten years old, he proclaimed he could provide the world with inexpensive and constant power, and he devised a "June bug engine" in which he glued four of the strongest beetles he could find to a tiny crosspiece of wood and had them spin a wheel, similar to mice on a treadmill. The bushes were black with these pests and they proved to be "remarkably efficient, for once they were started they had insufficient sense to stop." Unfortunately for Tesla and humanity, the town bully, the son of a retired army officer and whom the inventor referred to as an "urchin," devoured the bugs and destroyed the engine.

Electricity, however, elicited Tesla's most intense interest, and he claimed that his beloved cat, rather than school classes, introduced him to its wonders and mysteries. During a very cold and dry spell, he stroked Macak's back and it became "a sheet of light and my hand produced a shower of sparks." To Tesla, the static-electricity effect "was a miracle which made me speechless from amazement." The cat's body, he observed, became "surrounded by a halo like the aura of Saints!" His usually curious mother became alarmed, saying "Stop playing with the cat, he might start a fire." Such flashes prompted Tesla's childish imagination. "Is nature a gigantic cat and if so who strokes its back?" he asked. "It can only be God, I concluded." Decades later he added: "Day after day I asked myself what is electricity and found no answer. Eighty years have gone by since and I still ask the same question, unable to answer it."[34]

Teachers encouraged Tesla's newfound fascination. "I read all that I could find on the subject," he wrote, and "experimented with batteries and induction coils."[35]

Tesla's wondrous and inventive mind was haunted by the obsessive. According to him, trying to earn his father's respect prompted his "many strange likes, dislikes, and habits." Tesla went so far as to declare, "I would not touch the hair of other people except, perhaps, at the point of a revolver. I would get a fever by looking at a peach." He could not tolerate earrings on women; in his own words, the "sight of a pearl would almost give me a fit." Strangely, bracelets "pleased me more or less according to design." Tesla counted his steps when on a walk, making sure they were divisible by three—or he felt compelled to repeat the exercise. Such quirks persisted throughout Tesla's life. At each meal as an adult he demanded nine or eighteen napkins, a number divisible by three, and he wiped down each plate, bowl, and piece of silverware. Building on his mathematical talents, at the dinner table he also calculated "the cubical contents of soup plates, coffee cups and pieces of food—otherwise my meal was unenjoyable."[36]

Stranger still, the child born during a violent thunderstorm came to have episodes of seeing things with strobing intensity. This ordeal might have been enough to make anyone superstitious. Tesla described it as "a peculiar affliction due to the appearance of images, often accompanied by strong flashes of light that marred the sight of real objects and interfered with my thought and action." The images of objects became so confusing, he said, "I could not tell whether what I saw was real or not." Tesla even reached and passed his hand through an object, yet "the image would remain fixed in space."[37] These "tormenting" flashes accelerated during times of stress or exhilaration, and one particularly troubling image had "all the air around me filled with tongues of living flame."[38] (Later in life, this rationalist argued that these "tormenting appearances" were simply "the result of a reflex action from the brain on the retina under great excitation."[39] He also credited visions with allowing him to vividly see and manipulate inventions in his head.)

Tesla also labeled his sight and hearing "extraordinary." Able to "discern objects in the distance when others saw no trace of them," he boasted of saving several of his neighbors from fires by having heard "the faint crackling sounds which did not disturb their sleep" and calling for help. Later in life, he bragged about hearing "very distinctly thunderclaps at a distance of 550 miles," almost four times farther than any of his assistants.[40]

Tesla, however, admitted being a frail and vacillating child, having "neither the courage nor strength to form a firm resolve." His feelings surged between extremes, and he "was swayed by superstitious belief and lived in constant dread of spirits of evil, of ghosts and ogres and other unholy monsters of the dark."[41]

What some would no doubt treat with drugs, Tesla dosed with will-power; he was determined to control his thoughts and focus his mind. Believing the disturbing likenesses were not explainable by psychology or physiology, he concentrated on things he had seen in the real world. He also focused on images from books. Despite possessing a large library, his father discouraged him from reading, believing it strained his eyesight, and he "would fly into a rage when he caught me in the act." Yet each night the boy took to lighting hidden candles and covering his bedroom door's keyhole and cracks in order to enjoy isolated study.

Perhaps Tesla's greatest inspiration came when he discovered at the age of ten the Serbian translation of *Abafi* by Miklos Josika, a well-known Hungarian writer. Written in 1854, the story features a young nobleman, Oliver Abafi, who transitions through sheer determination from being disruptive and frivolous to being noble and disciplined, to the point where he sacrifices himself for his prince and country. The novel's hero motivated Tesla to utilize his own cerebral powers to control his feelings and actions. "At first my resolutions faded like snow in April," he wrote, "but in a little while I conquered my weakness and felt a pleasure I never knew before—that of doing as I willed. In the course of time this vigorous mental exercise became second nature. At the outset my wishes had to be subdued but gradually desire and will grew to be identical."[42]

Young Tesla's mental exercises led him to unexpected places. Admitting that he had accumulated limited experiences, he slowly discovered more satisfaction and peace by going "on in my vision farther and farther." "So I began to travel—of course, in my mind," Tesla admitted. "Every day, when alone, I would start on my journeys—see new places, cities and countries—live there, meet people and make friendships and acquaintances and, however unbelievable, it is a fact that they were just as dear to me as those in actual life and not a bit less intense in their manifestations." While on these journeys he "made imaginary friends, who were very dear to me and really seemed alive."[43] These images, of course, were no more real than the debilitating and random flashes that so troubled his youth, but he now controlled them. Such "incessant mental exertion," he later declared, "developed my powers of observation."[44]

These "journeys" also advanced his imagination. He "saw new scenes" that, initially, were "very blurred and indistinct," but through concentration they "gained in strength and distinctness and finally assumed the concreteness of real things."[45] Tesla concluded that this ability enabled him as a man to "visualize with the greatest facility."[46]

Tesla explored his imaginings in isolation. One aunt recalled when Tesla was with relatives "he would always like to be alone." Tesla went off in the morning "into the woods to meditate. He measured one tree to another, making notes, experimenting." Such solitary behaviors startled local peasants. According to a cousin, "They would approach and say, 'We're sorry; your [cousin] seems to be crazy.'"[47]

Tesla found nothing strange about concentrating attention upon himself. "It taught me to appreciate the inestimable value of introspection in the preservation of life," he reflected. He even criticized most people for being so "absorbed in the contemplation of the outside world" that they "are wholly oblivious to what is passing on within themselves."[48]

In addition to struggling with his thoughts, Tesla struggled to manage his health. One relative after another offered home cures to beef up his frail body. (Later in life, however, Tesla showed surprising agility and energy.) At the age of fourteen, in 1870, not long after completing his studies at the Gospic Gymnasium, Tesla became "prostrated with a dangerous illness or rather, a score of them," and he wrote, "my condition became so desperate that I was given up by physicians." To recuperate, he focused even more intently on reading, with the local librarian supplying scores of books, including the early works of Mark Twain (who became a good friend some twenty-five years later).

During this period, world events encouraged young Serbian men to seek new opportunities. As noted before, the Turks won a great battle on the Plains of Kosovo and dominated the Serbs for centuries. In 1869, the Austrian emperor added to their misery by retracting many of the scant privileges still enjoyed by Serbian peasants. As a result, many Serbs, including Tesla, began looking for a way out.

When Tesla regained his strength, Milutin sent him to Karlovac (Carlstadt) in order to prepare for the seminary and attend the Higher

Real Gymnasium, the nineteenth-century European equivalent of a middle school that prepared bright students for university. Tesla lived there for three years with Aunt Stanka, his father's sister, and Uncle Brankovic, a rugged colonel in the military, whom Tesla considered "an old war horse."[49] They offered the young student "an atmosphere of refinement and artistic taste quite unusual for those times and conditions,"[50] but they also required restraint. "I was fed like a canary bird," Tesla complained. "When the Colonel would put something substantial on my plate she would snatch it away and say excitedly to him: 'Be careful, Niko is very delicate.' I had a voracious appetite and suffered like Tantalus."[51]

Tesla, in fact, was prone to disease, and he contracted malaria in the low and marshy Karlovac, which is positioned at the confluence of four rivers. Despite drinking enormous quantities of quinine, he suffered for months.

Academically, Tesla, like his father and Uncle Josip, excelled at mathematics, and he completed the four-year curriculum in three years, graduating in 1873. He was particularly motivated by the school's physics instructor, Martin Sekulic, who designed his own models, including a radiometer featuring four tinfoil vanes that spun within a vacuum when struck by a bright light. The fascinated student declared, "It is impossible for me to convey an adequate idea of the intensity of feeling I experienced in witnessing his exhibitions of these mysterious phenomena. Every impression produced a thousand echoes in my mind. I wanted to know more of this wonderful force."[52]

Tesla, however, would not master freehand drawing, despite its being obligatory for engineering students. He imagined designs in his head, stubbornly refusing to put pen to paper. Being left-handed didn't help since drafting materials tended to be designed for righties. He found sketching to be "an annoyance I could not endure" even though he admitted his shortcomings and refusals were "a serious handicap . . . (that) threatened to spoil my whole career."[53] (Several years later, he willed himself to be ambidextrous, but he still disliked sketching complex designs, which prompted challenges to his patent applications.)

While at the Gymnasium, which he deemed a "fairly well equipped institution," Tesla directed his wandering imagination to create innovative—although not very useful—devices. Obsessed with building a flying machine that would "achieve what no other mortal had ever attempted," the immodest student theorized that a vacuum inside a cylinder would provide enough power to turn a shaft and propeller. He actually constructed a wooden box within which he arranged for air pressure to be "exerted at a tangent to the surface of the cylinder—a situation he knew would be required in order to produce rotation."[54] Tesla became "delirious with joy" when the shaft began to spin, but it wouldn't accelerate. Only after numerous adjustments did he realize "atmospheric pressure acted at right angles to the surface of the cylinder" and that the slight spinning resulted only from a small leak in the cylinder. "Tho this knowledge came gradually," he admitted, "it gave me a painful shock."[55] (The failure wasn't lost on him, though. This experience with air leaking into a vacuum and causing a small amount of motion prompted Tesla later to build a pioneering and efficient bladeless turbine.)

At seventeen, in 1873, Tesla "turned seriously to invention" with increased success. He had long practiced working things out in his head, owing to the rigors that his father imposed—performing elaborate calculations without putting pen to paper and memorizing long passages—and the rewards his father withheld. These mental exercises enabled him to "visualize with the greatest facility," and, unlike other engineering students, he claimed to need no drawings, models, or experiments. "When I get an idea," he wrote, "I start at once building it up in my imagination. I change the construction, make improvements, and operate the device entirely in my mind. . . . Invariably my device works as I conceived that it should, and the experiment comes out exactly as I planned it."[56]

Logic, or a deep trust in reasoning, underpinned Tesla's inventing. "There is scarcely a subject that cannot be mathematically treated," he maintained, "and the effects calculated or the results determined beforehand from the available theoretical and practical data."[57]

His growing thrill with engineering conflicted with his parents' persistent desire for him to join the priesthood. Here's another paradox in Tesla's early life—a father devoted to religion while his son loved science. Although Tesla admitted that "the mere thought of [joining the clergy] filled me with dread," he was "resigned to the inevitable with aching heart."[58] The stress of it weakened him further, and the fragile Tesla succumbed to the cholera epidemic sweeping through Gospic, where he had returned after his studies in Karlovac in 1873. For nine months, he fell "into dropsy, pulmonary trouble, and all sorts of diseases until finally my coffin was ordered."[59] He vomited for hours, lost dangerous amounts of fluid from persistent diarrhea, and revealed shriveled skin and sunken eyes. With severe muscle cramps, he could barely move; as he later explained, "My energy was completely exhausted, and for the second time I found myself at death's door." He was only seventeen years old.

And yet the bedridden Tesla mustered the spunk to suggest to his frantic father, "Perhaps I may get well if you will let me study engineering." Milutin pledged to his only remaining son, "You will go to the best technical institution in the world," to which Tesla later remarked, "I knew that he meant it. A heavy weight was lifted from my mind."[60] That relief, along with an herbal medicine that he described as "a bitter decoction of a peculiar bean," revived him "like another Lazarus to the utter amazement of everyone."[61]

Tesla, who could be dramatic, claimed numerous other youthful brushes with death, namely, to be "almost drowned a dozen times; was nearly boiled alive and just missed being cremated. I was entombed, lost and frozen. I had hair-breadth escapes from mad dogs, hogs, and other wild animals." He credited his survivals to being the result of "enchantment,"[62] and he gained a sense of invulnerability. The cholera episode, however, had a lasting effect, prompting Tesla to fear germs and avoid contact with others; he would not shake hands, let alone enjoy intimate relations.

As if disease and disaster were not enough, Tesla was also subject to bullies who invaded his private space. "There was a tough in our town," he explained, "who once thrashed a friend of mine because he parted his hair in the middle, and I henceforth resolved to do the same. I received likewise a thrashing which, for thoroughness, left nothing to be desired." The bully repeated these beatings for at least a year every time Tesla came home from school. Tesla, however, taught himself to box (which led to his later friendships with pugilists) and he eventually "had the satisfaction of returning the favors." They apparently sparred several times, yet just as Tesla "was about beginning to enjoy the experiences," he added in a macabre tone, "the fellow was shot in a brawl."[63]

Another obstacle was the army, which expected young men in its territories to serve for three years. Some accounts have Milutin deciding his now-recovered son should hide in the mountains, yet it's hard to imagine how Tesla, particularly with numerous military officers in his family, could avoid the service requirement, and there seems to be no official record of any deferment. Perhaps Milutin convinced his brothers to use their connections to have Tesla avoid conscription. The only comment Tesla made about this gap in his biography was that his father insisted he "spend a year undertaking healthy physical outdoor exercise," to which Tesla claims he "reluctantly consented." Whatever the reason, beginning in the early fall of 1874, Tesla disappeared, went into hiding, and spent nine months roaming the hills of Croatia "loaded with a hunter's outfit and a bundle of books."[64] He later admitted, "This contact with nature made me strong in body as well as in mind."[65] Such physical exercise contrasted with Tesla's early feebleness and almost exclusive focus on mental exercise.

The solitude also sparked some impractical scientific theories. Tesla came to believe, for instance, that he could use hydraulic pressure to shoot spherical containers of packages and letters through a pipe under the ocean. He didn't realize then that as the container's velocity increased so did the resistance it faced from the pipe walls. Another scheme he proposed was to improve passenger travel by placing giant rings around and above the earth and deploying enough reactionary forces to keep them stationary as the planet rotated. Travelers, he thought, could move up to a ring, jump on and watch the spinning planet rotate at about a thousand miles per hour, wait for their preferred destination, and then descend back to earth.

While Tesla wandered the mountains, his father secured for him a scholarship and acceptance to the Joanneum Polytechnic School, which Tesla considered to be "one of the oldest and best reputed institutions."[66] The young student headed off to Graz, Austria, in the fall of 1875, carrying a colorful shoulder bag his mother had embroidered and which he treasured the rest of his life.

Having dreamed for years of attending an engineering school, Tesla became a serious student, working from three in the morning to eleven at night, including Sundays and holidays. In addition to science, he tackled languages and voraciously read the classics, including Shakespeare, Goethe, and Spenser. Such efforts took him to the top of his class, where he passed through nine exams, nearly twice the requirement, and obtained numerous certificates. Tesla also developed "a verifiable mania for finishing whatever I began," which included reading the entire works of Voltaire, even though "to my dismay, there were close on one hundred large volumes in small print which that monster had written while drinking 72 cups of black coffee per day."[67]

Tesla's favorite lecturer was the physics professor Jacob Poeschl. Although considered "peculiar," as evidenced by his allegedly wearing the same coat for twenty years and because of his "enormous feet and hands," Poeschl demonstrated "perfection of his exposition." According to his eager student, "I never saw him miss a word or gesture, and his demonstrations and experiments came off with clocklike precision."[68]

This "methodical and thoroughly-grounded German" introduced Tesla to the serious study of electricity, as well as to the latest scientific equipment. In early 1877 the Polytechnic School acquired from Paris a Gramme generator, essentially a large horseshoe-shaped magnet that surrounded a hollow cylinder encased in tightly wrapped wire. This generator, when powered by a steam engine, enabled students to obtain reliable electricity rather than depend on expensive batteries or the erratic currents from early dynamos.

Building on the work of Hippolyte Fontaine of the Gramme

Company, Poeschl sought to demonstrate how electricity could be transmitted to where it was needed in order to run lights or motors. Connecting the generator to a motor, however, required painstaking adjustments of a commutator, a device that shifted electricity from its regular alternating state (similar to a sine curve in mathematics) into a single path of flow, or a direct current. The cumbersome commutator consisted of a metallic cylinder whose segments were insulated from each other and whose wire brushes, or stationary contacts on opposite sides of the cylinder, ensured the current from the generator moved in the same direction.

The commutator proved to be at the center of the direct-current versus alternating-current debate. Electricity usually is generated in bursts when a coil revolves around a magnet, passing the positive pole and then the negative pole. Those bursts switch directions very rapidly, first going clockwise, stopping, and then reversing themselves to go counterclockwise. Believing power needed to be in a direct continuous circuit (DC), early engineers devised a commutator to smooth out those bursts, allowing lights to avoid flickering and motors to run smoothly. Unfortunately, the commutator proved to be inefficient and prone to sparking, and it limited power distribution to only several hundred feet.

(A history and description of electricity are in an appendix at the back of this book.)

Tesla brashly decided problematic commutators needed to be eliminated, but the tradition-bound Poeschl declared flatly it could not be done. Tesla's favorite professor publicly rebuked him, saying that a commutator-less motor broke the laws of nature. "Mr. Tesla may accomplish great things, but he certainly never will do this. It would be equivalent to converting a steadily pulling force, like that of gravity, into a rotary effort. It is a perpetual motion scheme, an impossible idea."[69]

Tesla took Poeschl's criticism as a challenge, trusting his own instincts more than the professor's experience. "For a time I wavered," he wrote, impressed "by the professor's authority, but soon became convinced I was right and undertook the task with all the fire and boundless confidence of youth."[70] Believing our brains have some "finer fibers" that allowed us to "perceive truths which we could not attain through logical deductions," he vowed to devise a system whereby a motor, using alternating current, would operate efficiently with a generator, and he became obsessed with such an ideal, working on it "incessantly."[71]

While a sophomore at Joanneum Polytechnic, Tesla "had made up my mind to give my parents a surprise" of academic excellence, and his hard work and innate intelligence produced stellar results. Yet when Tesla went home at the end of the term expecting praise, Milutin criticized his son's accomplishments and work habits. The father's renewed rejection punctured Tesla's emerging self-confidence. (Only years later did Tesla discover a box of his father's papers that included letters from several professors recommending that he relax more and avoid overexertion. Milutin, of course, could have informed his son of such letters and encouraged him to take it easy, but he reverted to disapproval.)

Milutin's chastisement, according to Tesla, "almost killed my ambition."[72] The pupil, who had sought to be a professor of mathematics and physics, began to question the value of studying so hard. By his third year, beginning in the fall of 1877, Tesla took up gambling, a game of chance, rather than further his mathematics, at which he excelled. He stopped attending lectures and lost his scholarship. According to Tesla's roommate, "He began to stay late at the Botanical Garden, the students' favorite coffee house, playing cards, billiards and chess, attracting a large crowd to watch his skillful performances."[73] When they learned of his debauchery, Tesla's "cousins, who had been sending him money, therefore withdrew their aid." One day Tesla just disappeared, and "friends searched everywhere for him and decided that he had drowned in the [Mur] river."[74]

Without telling family or friends, Tesla had moved to Maribor in what is now Slovenia, about 185 miles from his parents in Graz. He worked during the day in a tool and die shop and spent evenings drinking and betting at the Happy Peasant pub. When a school friend finally located Tesla, he encouraged him to return home, yet Tesla nonchalantly responded, "I like it here; I work for an engineer, receive sixty forints a month, and can earn a little more for every completed project."[75]

Gambling, Tesla later admitted, became a "mania." "To sit down to a game of cards," he wrote, "was for me the quintessence of pleasure."[76] He and his friends played for "very high stakes," and "more than one of my companions gambled away the full value of his home." Despite his mathematical skills, Tesla claimed "my luck was generally bad."[77]

The schoolmate sent word to Tesla's parents, prompting Milutin to travel to Maribor in March 1879 and confront his son. Tesla refused to return home and arrogantly responded, "Is it worthwhile to give up that which I would purchase with the joys of Paradise?"[78]

The reactions of his parents differed substantially. Milutin, according to Tesla, "led an exemplary life, and could not excuse the senseless waste of time and money in which I indulged." The father gave "vent to his anger and contempt" and returned home shattered. Tesla's mother, however, took an alternative tact. "She understood the character of men," he wrote later, "and knew that one's salvation could only be brought about through his own efforts."

Although Tesla told his father he could "stop whenever I please," a few weeks later police arrested the gambling Tesla as a "vagrant" and deported him home to Gospic. The heartbroken Milutin, shocked that his remaining son had become a criminal, grew seriously ill and died at the age of sixty on April 17, 1879.

Tesla's only comment on his father's death was that the priest received a "funeral liturgy fit for a saint."[79] Still, although Tesla felt his father never appreciated him, Milutin's passing—particularly since Tesla's actions helped prompt it—must have been devastating.

Tesla stayed in Graz and returned to gambling until his mother intervened creatively. Djuka gave Tesla a roll of bills and said, "Go and enjoy yourself. The sooner you lose all we possess the better it will be. I know that you will get over it." Then she kissed her son. The strategy, a bit of a gamble itself, proved to be a conversion event. "I conquered my passion then and there," Tesla said. "I not only vanquished but tore it from my heart so as not to leave even a trace of desire."[80] (That is not exactly true. In fact, he enjoyed both cards and billiards when he later lived in Budapest and when he worked for Thomas Edison in the United

States; a colleague recalled: "He played a beautiful game. [Tesla] was not a high scorer, but his cushion shots displayed skill equal to that of a professional exponent of this art.")[81]

Tesla tempered several other habits and passions, too. Having been an obsessive smoker, burning fifteen or twenty big black cigars every day, he slowly realized the practice was damaging his health, so, according to Tesla, "my will asserted itself and I not only stopped but destroyed all inclination." He gave up coffee, worried that it was leading to heart trouble, although he confessed this particular discipline proved most difficult. He later admitted these denials preserved his life, but, more tellingly, he "derived an immense amount of satisfaction from what most men would consider privation and sacrifice."[82]

During this period of grief and restraint, Tesla briefly returned to church, inspired by the solitude and rituals of Serbian Orthodox services. His father had preached that the material universe possessed underlying laws—something he referred to as "logos"—that humans could discover. In fact, searching for such divine or ideal principles was a way to praise God. These lessons may have prompted Tesla to consider being an inventor as the best means of identifying that religious ideal as well as motivating him to find the ideal in his inventions.

At church, Tesla met the only person outside of his immediate family that he ever admitted loving. Anna was "tall and beautiful [with] extraordinary understandable eyes." The two enjoyed long strolls through town and along the river, where they exchanged stories and talked of a future in which he became an electrical engineer and she raised a family. Tesla admitted, "I fell in love." But the romance did not last, in large part because Tesla needed to "carry out my father's wish and continue my education."[83] Tesla left Anna and the church behind in January 1880 for the university in Prague; the two exchanged letters for a time, but their romance faltered and Anna soon married another man.

(Despite their breakup, Tesla and Anna kept in touch. Some twenty years later, when Tesla was thriving in New York City, she arranged for him to meet her son. The young man wanted to be a boxer, and Tesla, a fan of the sport, arranged for him to attend a boxing school near Madison Square Garden. Although Tesla tried to nurture the novice boxer, Anna's son rejected his advice and insisted on a tough opponent for his first fight. To Tesla's horror, the boy was quickly knocked unconscious and died shortly thereafter in a hospital. According to one account, "Tesla grieved for him as though he was his own son.")[84]

There are so many odd—and unexplained—paradoxes associated with Tesla's relationship to Anna. First, it's hard to imagine this germophobe being romantic. In fact, it's surprising this isolated and self-focused man could even fall in love with another person. Noting Milutin's persistent rejection of his son's engineering interests, it's also remarkable Tesla used his father's "wish" as an excuse to separate from the love of his life. Finally, you would have expected more words from Tesla about the death of the young boxer, who was the closest he would ever come to having a son.

When Tesla arrived at the Karl-Ferdinand University, Prague was part of Bohemia. He found the "old and interesting city was favorable to invention" in part because "hungry artists were plentiful and intelligent company could be found everywhere."[85] His mother had convinced her brothers, Petar and Pavle Mandic, to temporarily support Tesla's studies. He attended lectures that summer in philosophy, mathematics, and experimental physics. Since he did not want to study Greek or Czech, the required language classes, he participated as an unofficial auditor. Tesla devoted most of his time to considering designs for an electric motor at the Klementinum Library and the People's Café.

In order to curtail the motor's sparking, Tesla tried detaching the commutator from the motor and placing it on separate supports from the frame. Yet increasing the distance between the rotor and the commutator yielded no breakthrough. "Every day I imagined arrangements on this plan without result, but feeling that I was nearing a solution."[86]

Knowing his uncles' generosity was limited, Tesla "became a very fair example of high thinking and plain living," but he made up his mind to pursue his ideal motor.[87] When the local newspaper reported that

Thomas Edison was building a telephone exchange in Budapest, he asked his uncle Pavle to recommend him to Ferenc Puskas, who was supervising the Edison project and had served with Pavle in the Hussars, a light cavalry unit. In January 1882, Tesla moved to Budapest, the capital of Hungary, where he found both employment and inspiration.

2.

A GLORIOUS DREAM
Budapest

E lectricity fascinated and confused Tesla. Nineteenth-century physicists hypothesized that it moved mysteriously through ether, a medium that they thought permeated the atmosphere and outer space. Even now, when we understand the composition and properties of atoms, as well as how electrons behave, most scientists consider electricity a mysterious, versatile marvel that enables lighting, heating, motors, transport, communications, and computation.

Electricity is unique among the many material sources of power. We can see water rush over a wheel and turn paddles. Kerosene fills a lamp and ignites with a match. Natural gas is invisible but still a state of matter, which can be captured and measured out. In contrast, electricity must be generated; it is a physical effect—rather than a solid, liquid, or gas—occurring through the wires that conduct it.

Biographer Edmund Morris described electricity as "that fundamental force that lights our nights and floods our nanocircuits." It's "easy enough to see and feel what it does," he wrote, "but when electricity isn't going someplace, it has no color, no character."[1]

Since electricity does not exist naturally in quantities that can be manipulated for our benefit, producing and controlling this wonder has been an ongoing struggle. Inventors, engineers, entrepreneurs, and businesses have all gotten in on the act. But Tesla's advances, which began in Budapest, were key. Electrical power became the foundation for our modern society largely because of his inventions. (A more detailed review of electricity's history and properties can be found in the appendix.)

The Puskas brothers, friends of Tesla's uncle and successful European entrepreneurs, proved to be vital early supporters of both Edison and Tesla. Tivadar had met Thomas Edison a few years before at the inventor's laboratory. Tivadar impressed the Wizard of Menlo Park with his fancy car, wads of cash, and deep interest in the telegraph and telephone. The two agreed that the Puskas family, considered part of Transylvanian nobility, would promote Edison's phones, electric lights, and phonographs in Europe. The Puskas started in Paris by creating a telephone network linking hundreds of subscribers via a single switchboard, an improvement over the private lines connecting just two locations. They moved back home in late 1879 to do the same in Budapest.

Ferenc Puskas couldn't hire Tesla until he had arranged financing for the new project. So Tesla took a low-paying job with the Central Telegraph Office of the Hungarian government, where despite winning praise from the inspector-in-chief, he spent most of his time climbing poles to repair equipment, as well as doing routine drafting and calculations. Although Budapest was cultured and cosmopolitan compared to Smiljan or even Gospic, Tesla slept in a small apartment and, earning a meager salary of five dollars each week, lacked the resources to enjoy the city's café society. Still, he relished long walks along the Danube River and through Budapest's great parks with their fountains and lakes. Bored at work, however, he quit the telegraph job in order to concentrate on inventing, but the novice scientist did not attract any financial backers.

The stress of his own uncertain future caught up with Tesla in Budapest, where he "suffered a complete breakdown of the nerves." His fevered body shook and his pulse soared to 250 beats per minute. Perhaps hallucinating, he observed that "all the tissues of my body quivered with twitchings and tremors. A fly alighting on a table in the room would cause a dull thud in my ear. A carriage passing at a distance of a few miles fairly shook my whole body. The whistle of a locomotive twenty or thirty miles away made the bench or chair on which I sat vibrate so strongly that the pain became unbearable." Even in a dark room, Tesla claimed he "could detect the presence of an object at a distance of twelve feet by a peculiar creepy sensation on the forehead." A doctor "pronounced my malady unique and incurable," Tesla noted affectedly. "I clung desperately to life but never expected to recover."[2]

Later, Tesla regretted not being "under the observation of experts in physiology and psychology at that time." Although he claimed his hearing and sight were "always extraordinary," the emerging scientist believed doctors could have learned from his acute condition in Budapest.

Tesla usually ascribed any positive turns to his own remarkable will-power, yet he credited this recovery from deep depression to Anthony Szigeti, a former classmate and talented engineer who joined him in Budapest. Tesla proclaimed Szigeti "a remarkable specimen" whose body "might have served for a statue of Apollo." While Tesla's friend was not handsome—he had a lump on his head and a sallow complexion—he was "an athlete of extraordinary physical powers" who dragged the sickly Tesla through physical exercise. According to Tesla, "He saved my life."[3]

The two friends played billiards together and spent each evening eating at lively cafés and walking along the Danube River or through Varosliget, the major city park that featured a famous castle. They developed a friendly competition over who could drink the most milk, a contest Szigeti won after the thirty-eighth bottle.[4] According to Tesla, Szigeti gave him a "powerful desire to live and to continue the work."[5] Tesla's health slowly returned, as did "the vigor of my mind."[6] Tesla swore he would make a gradual but substantial transformation from being "so helpless a physical wreck" to being "a man of strength and tenacity."[7]

The recovered Tesla spent most of his waking hours focused on devising a commutator-less motor, trying to prove Professor Poeschl wrong. He reviewed articles and performed numerous calculations, but mostly he imagined. "I started by first picturing in my mind a direct-current machine, running it and following the changing flow of the currents in the rotor. Then I would imagine an alternator [an AC generator] and investigate the processes taking place in a similar manner. Next I would visualize systems comprising motors and generators and operate them in various ways. The images I saw were to me perfectly real and tangible."[8]

Once again, the work became an obsession. "With me it was a sacred vow, a question of life or death," he commented later. "I knew that I would perish if I failed."[9]

During one of his evening strolls with Szigeti in February 1882, Tesla experienced his eureka moment with respect to the electric motor. The inventor was reciting poetry, having the capacity to recall "entire books by heart, word for word."[10] As the sun set, he turned to Goethe's *Faust*.

> *The glow retreats, done is the day of toil;*
> *It yonder hastes, new fields of life exploring;*
> *Ah, that no wing can lift me from the soil*
> *Upon its track to follow, follow soaring!*
> *Then would I see eternal Evening gild*
> *The silent world beneath me glowing,*
> *On fire each mountain-peak, with peace each valley filled,*
> *The silver brook to golden rivers flowing.*
> *The mountain-chain, with all its gorges deep,*
> *Would then no more impede my godlike motion.*[11]

While Tesla uttered these stanzas, "the idea came like a flash of lightning and in an instant the truth was revealed." He grabbed a stick and quickly drew a diagram in the sand of an innovative motor that utilized a rotating magnetic field. "The images were wonderfully sharp and clear and had the solidity of metal and stone," he said, "so much so that I told (Szigeti): 'See my motor here; watch me reverse it.'"[12]

Years later the dramatic Tesla recounted the experience to the equally dramatic John O'Neill, the inventor's first biographer: "Isn't it beautiful? Isn't it sublime? Isn't it simple? I have solved the problem. Now I can die happy. But I must live. I must return to work and build the motor so I can give it to the world. No more will men be slaves to hard tasks. My motor will set them free. It will do the work of the world."[13]

Said with less flourish, Tesla had figured out how to create a motor that could utilize the undulating electrical rhythms of the more powerful alternating current. Discovering this rotating magnetic field could be considered the electrical correspondent of producing the wheel. A reporter with the *New York Herald Tribune* later explained Tesla's accomplishment as: "Up to this time everyone who tried to make an alternating current motor used a single circuit.... What Tesla did was to use two circuits, each one carrying the same frequency of alternating current, but in which the current waves were out of step with each other. This was the equivalent of adding to an engine a second cylinder.... [These alternating currents created] a rotating magnetic field [that] possessed the property of transferring wirelessly through space, by means of its lines of force, energy."[14]

Tesla described his rotating magnetic field as "a sort of magnetic cyclone which grips the rotable part [of the motor] and *whirls* it."[15] Late in life, he called this invention the closest one to his heart. "It was not only a valuable discovery, capable of extensive practical applications," he said. "It was a revelation of new forces and new phenomena unknown to science before."[16]

The maverick inventor relished upending standard practices. It was almost a quixotic quest for Tesla. While other engineers working on an alternating-current motor used a single circuit, Tesla utilized two circuits that generated dual currents ninety degrees out of phase with each other. When one current reached its maximum, the other had a zero value, creating a magnetic field that spun a piece of iron. Instead of churning up useless vibrations, Tesla's rotating magnetic field produced a turning force, as well as suggested revolutionary properties. Until Tesla, no one imagined a magnetic field could provoke rotation. As explained by a scientific historian, "The net effect was that a receiving magnet [or motor armature], by means of induction, would rotate in space and thereby continually attract a steady stream of electrons whether or not the charge was positive or negative."[17]

Rather than change the magnetic poles in the rotor, the part of the motor that rotates, Tesla also decided to change them in the stator, the motor's stationary component. As a separate science historian noted, "Tesla saw that if the magnetic field in the stator rotated, it would induce an opposing magnetic field in the rotor and thus cause the rotor to turn."[18]

At times in his life, Tesla described men, even himself, as automatons that simply react passively to the world around them. "I was but an automaton devoid of free will in thought and action," he wrote in 1919, "and merely responsive to the forces of the environment."[19] Yet that evening in Budapest, and on numerous other occasions, the proactive Tesla invented something completely new, or as he described it, he obtained insights by wrestling with nature "against all odds, and at the peril of my existence."[20]

Despite Tesla's bragging that he could visualize entire machines and systems that worked exactly as he planned, the alternating-current motor required years of testing and adjustments. As evidence that he might have practiced a bit of revisionist history when he later recalled the revelation in the park, Tesla did not mention the eureka moment when he applied for a patent, and Szigeti said nothing about the Budapest experience during his patent-defense testimony in 1889.[21]

Tesla's motor, which embodied totally new principles rather than refinements of existing work, represented a technological breakthrough that soon changed the world. It dramatically expanded the potential market for electricity, allowing it to be sold not just at night for lighting but also during the day for factories, appliances, and streetcar lines. For the first time, alternating current could be pumped hundreds of miles and efficiently power machines as well as lamps.

When demonstrating modesty, Tesla described his achievement simply: "A magnet, as it is well known, will attract a piece of wire and hold it. It was my good fortune to discover a method of constructing a magnet in such a way that it would not hold the iron, but spin it round and round."[22] Put another way, he figured out the difference between lifting and levitating an object.

With this innovation, Tesla had overcome the taunts from Jacob Poeschl, his former physics professor at the Polytechnic School in Graz, that a commutator-less motor was impossible. At only twenty-seven years old, his self-confidence soared, and he boasted: "I had carried out what I had undertaken and pictured myself achieving wealth and fame." More importantly, he demonstrated to himself that he had become an inventor, "the one thing I wanted to be."[23] "The inventor," he boasted, "gives to the world creations which are palpable, which live and work."[24]

Tesla in Budapest revealed a unique inventing style. Few discoverers quoted poets while drawing diagrams in the sand. Few approached creating so cerebrally. Inventing became Tesla's purpose and he perceived it as noble. "The progressive development of man is vitally dependent on invention," he wrote. "It is the most important product of his creative brain. Its ultimate purpose is the complete mastery of mind over the material world, the harnessing of the forces of nature to human needs."[25]

A newly self-assured Tesla basked in the "intense enjoyment of picturing machines and devising new forms." Visions came at him in an uninterrupted stream and he experienced "a mental state of happiness as complete as I have ever known in life." The cerebral engineering—when "the only difficulty I had was to hold [the ideas] fast"—allowed him to conceive of apparatuses, including dynamos and transformers, that were to him "absolutely real and tangible in every detail, even to the minutest marks and signs of wear."[26] He even envisioned his motor operating on three or more alternating currents simultaneously, what he described as a poly-phase system.

Still, Tesla's visions needed to be converted into working devices, so the inventor decided to visit Ganz and Company. This Budapest factory, which had begun as an iron foundry making wheels for railway cars, had recently converted to constructing electrical equipment that powered arc and incandescent lights. There the young Serb became intrigued with a ring transformer—essentially a large iron loop encircled by two coils—that had been discarded in a corner of the workshop. Perhaps playfully or maybe with insight, Tesla placed a ball atop the device—and it began to wobble. He concluded the windings in the two coils produced two alternating currents, which created a rotating magnetic field that turned the ball, thus confirming his visualization for an alternating-current motor capable of performing all kinds of work.

Tesla was not the only scientist investigating rotating magnetic fields. Three years earlier, Walter Baily had demonstrated the principle before the Physical Society of London, and Galileo Ferraris had designed an AC system in Turin, Italy. Tesla actually acknowledged it was "not new to produce the rotation of a motor by intermittently changing the poles of one of its elements," but he alone discovered "the model or method of utilizing such currents."[27] A U.S. judge eventually declared Baily's ideas to be "impractical abstractions,"[28] and Professor Ferraris admitted, "Tesla developed it much further than [I] did."[29]

Tesla's need for income eventually overwhelmed his glorious but uncompensated visions, so he joined Ferenc Puskas when construction began of the Edison telephone exchange in Budapest. The young inventor worked diligently, even developing an innovative conical-shaped amplifier or repeater, a precursor of the loudspeaker. Tesla never patented the ingenious device, although he later bragged that its originality compared to his better-known inventions. Once the exchange became operational, Ferenc sold it, and his brother, Tivader, invited Tesla and Szigeti to come to Paris in late 1882 to help install an Edison electric lighting system.

The previous year, Edison had amazed cosmopolitan Parisians with a light display at the Electrical Exposition. He installed within the vast exhibit hall a working lighthouse, jumbo generators, and a brilliant collection of five hundred incandescent lamps of sixteen candlepower each. About electricity, one baffled Frenchman wrote: "We are not yet in the habit of observing machines that function without apparent cause. Their occult workings baffle us. The secret of their existence escapes us."[30] Although confused by electricity's actual workings, the impressed French conferred the Legion d'Honneur on Edison.

Tesla enjoyed his time in Paris, a city he called "magic." He spent his first days roaming "through the streets in utter bewilderment of the new spectacle." The dining and drinking attractions, he said, "were many and irresistible," but the income, he lamented, "was spent as soon as received." In fact, when Puskas asked how he was getting along, Tesla replied, "The last twenty-nine days of the month are the toughest!"[31]

The young Serbian engineer rented a room on Boulevard St. Marcel, on the edge of the picturesque Latin Quarter filled with students and professors. Tesla relished his morning routines, swimming twenty-seven laps (divisible by three) at a bathhouse on the Seine, devouring "a woodchopper's breakfast" at half past seven o'clock, and walking a half hour to the southeast in order to reach his work at the Ivry-sur-Seine lamp factory recently built by Compagnie Continentale Edison. His evenings were devoted to wandering the gaslit avenues, enjoying the new electric lighting that he helped install at the gilded Paris Opera, and playing billiards with his co-workers.[32] He also savored Parisian nightlife; one evening the famous actress Sarah Bernhardt dropped her scarf near his dining table, which prompted an intense and flirty conversation. The two subsequently met on a few occasions in Paris and New York, although the handsome inventor claimed the relationship was platonic. Still, Tesla saved Bernhardt's scarf, without washing it, his entire life.

Once again drawing in the dirt with a stick, Tesla showed some Edison men his vision for an AC motor that utilized several alternating currents. "My idea," he said, "was that the more wires I used the more perfect would be the action of the motor."[33] Yet the pragmatic and cost-conscious Edison engineers scoffed at the scientist's ideal, complaining copper wires were the most expensive part of an electrical distribution system; they wanted less, not more, wiring.

Still, the Paris assignment allowed Tesla to gain practical experience with existing designs for dynamos and motors. The Edison Company, meanwhile, profited from Tesla's mathematical training and his ability to complete complex calculations, such as the best length and diameter for a motor's coils. Tesla developed an innovative automatic regulator that enhanced the performance of Edison's dynamo, and he traveled to install lighting systems at Berlin cafés and a Bavarian theatre. He also advanced arc lighting along the Avenue de l'Opéra and other streets, helping to transform Paris into the City of Light.

Viewing the young scientist as their new problem solver, Edison executives sent him in February 1883 to the Strasbourg railway station. During a dedication ceremony with German Emperor Wilhelm I, the lighting system had short-circuited and exploded. With his ability to speak German and French, Tesla tackled the complex setup of four direct-current generators and 1,200 incandescent lamps, as well as five alternating-current generators and sixty arc lights. He determined the problem to be defective wiring within the underground conduits, which were a relatively new way to string electric lines.

While in Strasbourg for almost twelve months, Tesla and Szigeti made time at night to focus on their AC motor, working out of a closet in order to ensure secrecy. They were after, as Tesla stated, "the simplest motor I could conceive of... [having] only one circuit and no windings on the armature or the fields."[34] When their initial design did not work, the two engineers began fiddling. They tried iron and then steel instead of brass within the stator's core, and then they positioned by trial and error a steel bar so its magnetic field repelled the induced currents and caused the disk to rotate. By the summer's end, a satisfied Tesla boasted to have "finally had the satisfaction of seeing rotation effected by alternating currents of different phase, and without sliding contacts or commutator—all as I had conceived a year before."[35]

Tesla consistently claimed to complete his inventing in his brain. Later in life, he explained his creative process: "In my work, I first get a 'feeling' that there is a solution to a problem. This feeling means a great deal to me, for when I get it I always succeed eventually in working out the problem. Probably it is an indication that the subconscious mind is diligently on the track or has already found a solution. Then I think generally over the problems, not concentrating on any one point. I may go for months, or even years, with an idea in such a state in the back of my head. Finally I close around on the idea, and the image that was at first blurred gets sharper and sharper—until in time it becomes a reality. After the general form is ascertained, I proceed to perfect the machine or idea—decide whether one thing or another will improve it. The vision is so clear that I can tell if a certain part is out of balance! And finally I can give exact measurements to the workmen without having made even a sketch. The models are not to experiment with, but are built simply to prove the ideas were created in my mind."[36]

That's an exaggeration; Tesla spent hours tinkering with different models of motors and later admitted his initial ideas often needed adjustments before they became practical devices. Still, his inventing style was unique, a product of his vivid imagination and powers of memorization, and it differed substantially from Thomas Edison's empirical process. To obtain pleasing indoor illumination, unlike the blazing and ugly glare from arc lights, Edison tested hundreds of filaments, from cotton to platinum, before settling on an incandescent bulb composed of a thin wire of carbon within a vacuum tube.

Tesla tried to secure financing in Strasbourg to develop his AC motor. He attracted the attention of a former mayor who approached several wealthy businessmen, but in the end, no one was willing to finance an unproven inventor and a revolutionary design whose operation and value they did not understand. The best Tesla obtained was a fine bottle of St. Estephe wine of 1801, which the mayor had buried when the Germans had invaded the country in 1870.

Tesla also obtained little from the Edison Company, which he believed had promised him a bonus if he solved the Strasbourg train station mystery. What he gained was an introduction to corporate bureaucracy, where one administrator after another suggested someone else was the decider. "After several laps of this *circulus vivios*," a cynical Tesla finally concluded, "it dawned upon me that my reward was a castle in Spain."[37]

Charles Batchelor—Thomas Edison's closest advisor, a master electrician, and director of the Edison companies in France—subsequently invited the inventor to travel to New York, meet the Wizard of Menlo Park, and improve the dynamos being developed at the Edison Machine

Works. Without many European options to advance his alternating-current research, Tesla decided to try his fortunes in "the land of Golden Promise."[38] Tivadar Puskas encouraged the transition and addressed a letter to his friend Edison, stating, "I know two great men and you are one of them; the other is this young man."[39]

3.

REVERBERATION OF
HEAVEN'S ARTILLERY
New York

Tesla's journey to America proved rough. He lost his wallet and ticket on his way to the dock and needed to recite from memory the voucher's eleven-digit number, berth location, and even the first mate's name in order for the agent to allow him aboard. The voyage itself presented "a mutiny in which I nearly lost my life"[1] as burly sailors protested filthy beds, the lack of clean water, and roaming rats. Tesla, however, spent most of his time "at the stern of the ship watching for an opportunity to save somebody from a watery grave."[2]

Tesla arrived in New York City in early June 1884 on the *City of Richmond*. He carried little with him, only "remnants of my belongings, some poems and articles I had written, and a package of calculations relating to solutions of an unsolvable integral and to my flying machine."[3] He claimed his financial resources totaled only four cents.

When Tesla set foot in Manhattan, the Statue of Liberty was under construction and Thomas Edison's first electric generating plant had opened on Pearl Street two years earlier. Like thousands of other immigrants—Germans, Italians, Irish, and Scandinavians—Tesla arrived with dreams. The United States, however, did not present the young immigrant with positive first impressions. Before Ellis Island opened in 1892, immigrants came into Manhattan's Castle Clinton depot, a gloomy former fort at the southern tip of Manhattan Island. One customs officer misunderstood the new arrival and recorded Tesla's place of birth as Sweden (rather than Smiljan), while another immigration official barked "Kiss the Bible. Twenty cents!"[4] A burly policeman, perhaps angry at the growing wave of Eastern European immigrants, twirled his stick, which "looked to me as big as a log," and delivered directions "with murder in his eye." Tesla recalled asking himself, "Is this America? . . . It is a century behind Europe in civilization."[5] Noting the absence of broad boulevards and public gardens, he further observed: "What I had left was beautiful, artistic, and fascinating in every way; what I saw here was machined, rough, and unattractive."

Yet America offered the new immigrant opportunity in the first few blocks. While walking uptown on that pleasant Friday, Tesla encountered the foreman of a small machine shop who was trying to repair "an electric machine of European make." The man had given up the task as hopeless, but Tesla undertook the challenge, stating, "It was not easy, but I finally had it in perfect running condition." Although not expecting any compensation, Tesla claimed to have received from the foreman twenty dollars, a substantial sum (the equivalent of more than five hundred dollars today) compared to the four pennies in his pocket. He expressed his wish "that I had come to America years before."[6] Five years later, he declared the United States "was more than one hundred years *ahead* of Europe."[7]

Tesla had been working for several years for Edison's companies, and on his second day in the United States, the twenty-eight-year-old immigrant actually met thirty-seven-year-old Thomas Edison. What Tesla described as "a memorable event in my life" was also an immediate clash of personalities and cultures.[8] Although essentially broke, Tesla wore an immaculate bowler hat and white gloves. With a slim build and deep-set eyes, he presented himself as a worldly gentleman who spoke several languages fluently. Edison, in contrast, was frumpy and grumpy, about five inches shorter, and his English, muddled by chewing tobacco, revealed a Midwestern drawl.

The Edison headquarters at 65 Fifth Avenue reflected the young company's confidence and practicality. Located in an impressive brownstone, the main floor featured sparkling electric chandeliers and lamps to impress potential investors. Edison spent most of his time in a comfortable back office, reviewing patent applications and smoking cigars. Since the expanding firm needed trained men, one floor served as a night school for fledgling engineers, while unmarried employees actually lived on the top floor.

As a youngster, Edison appeared destined to be neither prolific nor famous. He descended from a line of rebels. His grandfather, a prosperous Tory, fought against George Washington, was convicted of treason, was sentenced to hang, and fled to Canada. His father also narrowly escaped, this time from Canada to Michigan after participating in an unsuccessful coup against the Royal Canadian government.

Thomas Alva Edison was born in February 1847 in Milan, Ohio, the family's seventh and final child. Al, as he was called, spent his early years dreaming, drifting, and getting into trouble. Although possessing an encyclopedic memory and vivid imagination, Edison performed poorly at his one-room school, with a teacher describing the student as "a little addled." His mother eventually homeschooled Al and nurtured his love of reading and science. "My mother was the making of me," Edison declared later, using words similar to Tesla's for his own mother. "She understood me. She let me follow my bent."[9]

One of Edison's first experiments created a spark that burned his father's barn to the ground, yet his punishment in the public square failed to deter his pranks. He also shocked any friend or relative gullible enough to touch his electric generator.

In his teens, Edison moved with his family to Port Huron, Michigan, where he proved to be a decent telegraph transmitter and adept tinkerer. At the age of twenty-four, he settled in Newark, New Jersey, where, with growing confidence in his own abilities, his dogged dabbling eventually produced useful products that brought him to the attention of industrialists and Wall Street financiers. His stock ticker, for instance, overcame many of the telegraph industry's bottlenecks by operating at two hundred to three hundred words per minute, and his automatic duplex allowed two messages to be sent simultaneously on a single wire. His eventual list of discoveries reads like a litany of modern products: phonograph, motion picture camera, incandescent bulb, parallel circuit, and mimeograph machine.

Unlike the isolated Tesla, Edison worked best when he cooperated with a team, yet he remained the lab's driving force, whose persistence prompted continual experimentation. He overflowed with ideas, yet his successes resulted more from trial and error than from envisioning finished products in his mind. The Wizard of Menlo Park fiddled, built scores of models, and tested hundreds of materials. As Edison himself put it: "Genius is 1 percent inspiration and 99 percent perspiration."[10]

The development of the incandescent lamp is illustrative. As noted briefly above, Edison tried almost every imaginable chemical (e.g., chromium, molybdenum, boron, silicon, and zirconium oxide) to coat almost every imaginable substance (e.g., fish lines, cotton, cardboard, wood shavings, bamboo, and beards). He initially dismissed carbon as a possible "burner" because of its presumed weakness when exposed to the 3,000-degree Fahrenheit heat of an electric current, but he decided to bake carbonized sewing thread and to wire the charred ribbon to a stem assembly within a globe. After his new pump exhausted the bulb of its air, he turned on the electric current. The first eight attempts produced only a broken thread, but on October 21, 1879, a slightly thicker thread burned for forty straight hours. It might have lasted longer, yet Edison, ever the tinkerer, tested higher voltages until the filament expired.

Edison even rejected Tesla's trust in mathematics, bragging about not appreciating scientific theories. "At the time I experimented on the incandescent lamp I did not understand Ohm's law," admitted Edison. "Moreover, I do not want to understand Ohm's law. It would prevent me from experimenting."[11] (Understanding Ohm's law, however, would have revealed a core problem with Edison's direct-current approach, since the principle asserts that the amount of electricity delivered, the current, is equal to voltage divided by resistance; in other words, direct-current electricity—with its low voltages—requires heavy and expensive copper wire, faces resistance, can travel only short distances, and is, therefore, limited in its applications to small, high-density areas.)

While Tesla loved inventing for its own sake, Edison was first and foremost an entrepreneur. There was a clear purpose behind his experiments—to make money. He understood and could navigate within the world of finance. In 1890, he and Henry Villard, a savvy investor and the force behind the Union Pacific Railroad, reorganized and capitalized at twelve million dollars the innovator's various companies into Edison General Electric.

Despite their different styles and motivations, the two inventors shared several characteristics, including their capacity for showmanship and boastfulness. On September 4, 1892, Edison scheduled the debut of his Pearl Street generating station, which he cleverly located near the stock exchange, major banking and financing houses, and leading newspapers. He knew that a successful demonstration would prompt brokers, bankers, and editors to advance his financial success and fame. At precisely 3:00 p.m., an electrician threw the switch that fed current from a Jumbo generator (named after the great elephant brought to America by P. T. Barnum) to 106 lamps throughout Morgan's office. Fifty-two additional bulbs glowed in the *New York Times'* editorial office. The next day's newspaper described the artificial light as "soft and mellow to the eye; it seemed almost like writing by daylight."[12] Edison, obviously pleased with his performance, declared, "I've accomplished all I promised."[13]

Both electrical wizards attracted skeptics as well. Professor Henry Morton of the Stevens Institute labeled Edison's lamp "a conspicuous failure, trumpeted as a wonderful success. A fraud upon the public." The *London Times' Sunday Review* suggested Edison's results were based on trickery and declared, "There is a strong flavor of humbug about the whole matter."[14]

The Menlo Park Wizard also had to battle the legal challenges of his competitors. In part because Edison's early applications were slipshod and chaotically drawn, patent commissioners in October 1883 ruled that William Sawyer's incandescent lamp with a carbon filament preceded Edison's. They granted Sawyer the legal monopoly to sell the lamp for the next seventeen years. Hiram Maxim, already the producer of the first portable fully automatic machine gun, tried to strengthen the fragile carbon filament by burning it within a Sawyer bulb filled with hydrocarbon vapor. Commenting on this "flashing process," *Illustrated Science News* predicted: "In connection with electric illumination [Maxim's] name will be remembered long after that of his boastful rival [Edison] is forgotten."[15]

The pride and doggedness enabled both Tesla and Edison to triumph over the naysayers and lawyers. Over the years, the two sometimes worked together and often fought each other, but their stories and struggles reveal much about the making of our modern economy.

Notwithstanding their awkward initial exchanges, Edison hired Tesla that same afternoon to fix a lighting system aboard the SS *Oregon*, then the world's fastest passenger liner and the first to use electric lighting. Tesla worked through the night to repair the short circuits and breaks in the ship's dynamo. When walking home at five o'clock in the morning, he ran into Edison and Batchelor on Fifth Avenue, also leaving their work. The surprised Wizard of Menlo Park listened silently to his new employee's update, but, according to Tesla, "when he walked some distance away I heard him remark: 'Batchelor, this is a damn good man,' and from that time on I had full freedom in directing the work."[16] Edison quickly offered Tesla a job at his Machine Works for eighteen dollars a week, where Tesla fixed incandescent and arc lamps and adjusted direct-current motors. In fact, Edison got far more than a talented mechanic. Within six months, Tesla had designed two dozen machines that replaced the less efficient versions used by the Edison companies, saving the older inventor enormous sums of money.

As it happened, Tesla shared Edison's drive; they both thrived on work and needed little rest. Edison was known to toil for days, taking only occasional catnaps on his office sofa or a tabletop, while Tesla's working hours in the United States were ritualized and long, beginning at 10:00 a.m. one morning and going until 5:00 a.m. the next morning. (Tesla, even in old age, slept only two to three hours a night.) Edison said of Tesla, "I have had many hard-working assistants, but you take the cake."[17]

The two inventors, however, were not equally sophisticated. Edison, in fact, could be downright crass. When, for instance, he could not locate Tesla's small village on a map, he assumed the Serb had grown up in an uncivilized land and asked, "Have you ever eaten human flesh?"[18] The polished and reserved Tesla was no doubt nonplussed.

Tesla nonetheless worked tirelessly on Edison's behalf. He tried to convince Edison of AC's advantages, suggesting it liberated the Edison systems from the shackles of their one-mile-radius limitation. According to Tesla, Edison responded "very bluntly that he was not interested in alternating current; there was no future to it and anyone who dabbled in that field was wasting his time; and besides, it was a deadly current whereas direct current was safe."[19]

The major rift between the inventors came after Tesla believed he'd been cheated rather than overworked. When Tesla insisted that he could dramatically increase the efficiency of Edison's dynamos, his boss expressed skepticism, and then, according to Tesla, promised him fifty thousand dollars for success. That was a large amount of money considering Edison's reputation for stinginess and his company's own dire need for cash and capital. Tesla worked for six months until he discovered that shorter-core magnets yielded far more energy and tripled the Mary-Ann dynamo's output. Tesla had made good on his boast, but when he entered Edison's smoke-filled and cluttered office in early December 1884 asking to be paid, the Wizard fell forward in laughter and declared his monetary offer had been made in jest: "You are still a Parisian! When you become a full-fledged American you will appreciate an American joke."

No doubt Tesla should have recognized Edison's exaggerations, yet he was stung by his hero laughing in his face, essentially mocking his accomplishment and making fun of immigrants. Tesla, in fact, had fixed Edison's mistakes at the Strasbourg train station and on the SS *Oregon*, and he had vastly improved Edison's machines. Despite the boss's subsequent offer of a raise of ten dollars per week (a 55 percent increase), Tesla resigned, picked up his bowler hat, and walked out. That evening Tesla scrawled in his notebook: "Good by [sic] to the Edison Machine Works!"[20]

Not long after leaving Edison's workshop, Tesla received a new offer. Benjamin Vail, a town council member in Rahway, New Jersey, had long been fascinated with electric lighting and wanted his village to be known for its modern technologies. He committed one thousand dollars and raised another four thousand dollars from local businessmen to launch the Tesla Electric Light and Manufacturing Company in late December 1884. Tesla moved into an apartment in Lower Manhattan, where children regularly stole the colored glass balls he had placed on decorative sticks throughout his garden,[21] and he opened a small laboratory on Liberty Street that was surrounded by a bustling wholesale food market.

The scientist, expanding on the arc-lighting work he had done for Edison, toiled for a year of "day and night application" to improve both generators and lamps, for which he obtained several patents that he assigned to the company in exchange for stock shares. By August 1886, the completed system lit Rahway's streets and powered its workshops. Rahway was featured on the front page of *Electrical Review*, a trade journal, and the Tesla Electric Light and Manufacturing Company advertised its system as "the most perfect Automatic, Self-regulating system of electric arc lighting yet produced . . . with no flickering or hissing."[22]

Ultimately, Benjamin Vail, like Edison, took advantage of the inventor. Vail lost interest in manufacturing arc-lighting equipment, as that side of the operation became dominated by the Brush and Thomson-Houston companies. The executive instead preferred to focus on operating the Rahway system, which didn't need Tesla's expertise or inventions. As the company controlled Tesla's patents, the scientist received no royalties, only "a beautifully engraved certificate of stock of hypothetical value."[23] Tesla described the rejection as "the hardest blow I ever received. . . . I was forced out of the company, losing not only all my interest but also my reputation as an engineer and inventor."[24] It would not be the last time Tesla was bested by a businessman.

Virtually broke and without engineering job prospects, Tesla "lived through a year of terrible heartaches and bitter tears, my suffering being intensified by material want." For many of those days he did not know "where my next meal was coming from," yet, "never afraid to work," he approached a crew digging ditches for underground telephone and telegraph lines. Although the foreman initially laughed at his good clothes and white hands, Tesla claimed later, "I worked harder than anybody. At the end of the day, I had $2."[25]

While Tesla (and thousands of other unemployed men) struggled during the economic recession of 1887–1888, Edison's expanding company battled its own employees wanting better pay and conditions. At its New Jersey lamp factory, for instance, eight filament workers joined a union and, according to Edison, "became very insolent, knowing that it was very impossible to manufacture lamps without them." The Menlo Park inventor responded by designing thirty machines that automated their tasks and then bragged: "The union went out. It has been out ever since."[26]

Even in this dark period, Tesla could not stop imagining new machines. He completed a patent application for a thermomagnetic motor that took advantage of the fact that iron magnets lose their magnetism when heated. Tesla's motor featured a pivot arm that moved in and out of a flame, causing a flywheel to spin. The machine itself never proved to be profitable, but Tesla's ditch-digging foreman became impressed by the young man's ingenuity and introduced Tesla to Alfred Brown, superintendent of Western Union's New York Metropolitan District. Brown was considered a "first class electrician and expert in underground telegraph work." He introduced Tesla to Charles Peck, a patent lawyer from Englewood, New Jersey, and secretary of the Mutual Union Telegraph Company, which supplied and operated twenty-five thousand dedicated telecommunication lines in twenty-two states for banks wanting their data exchanges secured.

Peck and Brown offered the business acumen and temperament Tesla so desperately needed. They ensured Tesla was grounded in reality, giving him license to imagine and invent. They understood how to finance companies, promote innovative technologies, and position inventions to receive the most publicity. Importantly, they also were honest and appreciated Tesla's talents. According to Tesla, "They were in all their dealings with me the finest and noblest characters I have ever met in my life."[27]

Tesla, Peck, and Brown in April 1887 formed the Tesla Electric Company, with the inventor to receive a third of the profits, Peck and Brown to split a third, and the final portion to be invested in developing new innovations. Tesla, moreover, obtained a monthly salary of $250 (the equivalent of about $78,000 annually today) as well as access to a slightly larger laboratory in Manhattan's financial district at 33–35 South Fifth Avenue (now West Broadway), not far from Edison's downtown office. Although furnished with only a stove, workbench, and dynamo, he relished his new engagement and struck a deal with the Globe Stationery & Printing Company, located on the ground floor, to obtain power at night after Globe turned off its presses. Tesla and Szigeti, who landed in New York City in May 1887, also renewed their habits of strolling through parks, reciting poems, and discussing design options.

Tesla promised to consider a variety of projects, not just an AC motor, and the new firm quickly developed a commutator that reduced a dynamo's sparking, as well as a short-circuiting commutator that increased a DC motor's efficiency. Tesla, however, focused most of his attention for almost a year on a pyro-magnetic generator that aimed to convert the heat from burning coal into electricity with less waste. Tesla tried to alternately heat and then cool a magnet in order to directly induce a current in a conductor. This innovation bypassed the usual method, which was having coal-fired heat boil water to produce steam to power an engine that spun the same dynamo, each step of which wastes energy. Unfortunately, the inventor could not achieve sufficient temperature differentials to provoke a current, and the patent office rejected his application.

Tesla worried that Brown and Peck might abandon him, just as Vail had done in Rahway. Yet Peck, the lawyer, showed faith in the inventor and said comfortingly, "Now do not be discouraged that this great invention of yours is not panning out right; you may bring it to a success after all. Perhaps it would be good if you would switch to some other of your ideas and drop this for a while. I have had an experience that this is a very good plan."[28]

Tesla shifted his attention back to his favorite idea from Budapest—an AC motor powered by a rotating magnetic field—that he believed would help solve two challenges facing electricity's adoption. First, early power distribution presented a dangerous and jumbled mess. Brooklyn residents became so familiar with skirting the sparks from electric trolley tracks that they named their baseball team the Dodgers. Well before common standards, different wires transmitted power at different frequencies; moving across the street often required purchasing new lights and appliances because that area might be controlled by another electricity provider. A chaotic tangle of poles and crisscrossing wires blocked the sky; some of those lines lost their insulation, releasing fiery discharges and even electrocuting several careless children playing under the poles. Some uniformity was needed if electricity was to attract customers and gain market share.

Second, some twenty different electrical "systems" operated in the United States, each with its own limitations. Charles Brush's high-voltage, direct-current approach could supply power to arc lights for street lighting, but those lamps produced a ghastly white glare and buzzed disagreeably; that system couldn't run indoor lamps or appliance motors, nor could its power be sent any significant distance. Thomas Edison's low-voltage, direct-current scheme lit incandescent lamps in densely populated New York neighborhoods, but it only distributed electricity short distances and without enough force to power large machines. George Westinghouse planned for a high-voltage, alternating-current system that could send power over long distances, but no motors ran on AC. Each approach, in short, was designed to meet a specific need. None of them supplied electricity in substantial amounts for multiple uses.

Tesla's genius was to provide order to the electricity business and to create an efficient motor that allowed alternating current to meet numerous demands. His inventions helped bring needed uniformity to an emerging power industry. While other engineers tried to devise better commutators, Tesla envisioned a totally new approach that linked alternating currents that were out of phase with each other. According to Tesla, this "poly-phase" system overcame the limitations of his competitors.

With Brown and Peck offering the time and resources to convert his idea for an AC motor into a prototype, Tesla initially tried to modify the Weston dynamo in his workshop, experimenting with different windings. He eventually focused on sending separate currents to separate coils, and then he ensured those currents were out of phase with each other, meaning that when one was at its positive-value peak the other reached its maximum negative value. Not dissimilar to the wobbling ball atop an old ring transformer at Ganz and Company about five years before, Tesla now placed a shoe-polish tin on a pin in the center of his own innovative ring. The tin began to spin rapidly. Success!

It was still a long way from a spinning tin can to a commercial AC motor. Tesla also understood that a workable motor depended on an entire system, including a transformer that raised the voltage (the electric pressure in a line) and sent power long distances, and then another transformer that reduced that voltage before the current entered a home, allowing the power to illuminate lamps and run appliance motors.

Such transformers also intrigued Westinghouse, who had already proved his technical genius by developing air brakes for railroad cars. A skilled businessman, the Pittsburgh-based Westinghouse foresaw a profit opportunity for alternating-current systems to challenge Edison's dominance in the electricity industry. Because of DC power's range restrictions, Edison's markets were limited to densely populated areas. However, the Edison system had the advantage of being the entrenched approach and was perceived as being the safest. Westinghouse argued AC systems could profitably serve more cities and towns, if effective transformers and motors could be developed. This dispute between direct current and alternating current—between Edison and Westinghouse/Tesla—became known as the "War of the Currents."

The stakes were huge. The victor would reap enormous riches by controlling the infrastructure and intellectual property of what was to become the nation's largest industry. Yet the contest promised far more than personal wealth: whoever won would determine the speed and breadth of a new economic and cultural age. Electricity promised to power emerging industries as well as new appliances that could keep food fresh and buildings cool. The length of working days would no longer be determined by the sun. Elevators and streetcars would raise and expand the urban landscape. Electric-powered street lamps even promised to reduce violent crime. Achieving those opportunities certainly required inventors, but also lawyers, publicists, and investors.

While Peck and Brown certainly hoped Tesla's alternating-current innovations would win out, they initially pushed for an improved DC device; with the prevalence of DC systems in the United States, that innovation would find a ready market. Stubbornly clinging to his AC vision, Tesla orchestrated a dramatic demonstration to change their priorities and open up the wallets of investors. He invoked the story of Christopher Columbus challenging his critics to balance an egg on its end; when Columbus thought outside the box and cracked the egg's end slightly, so the tale goes, an impressed Queen Isabella invested in the explorer's transatlantic project. Similarly, Tesla placed a copper egg on top of a wooden table, beneath which he had placed a four-coil magnet producing alternating current. When the scientist flipped a switch, the egg began to spin, and as its rotation accelerated, the egg stopped wobbling and stood on its end, revealing both motion and stasis at the same time. The astonished Peck and Brown became fervent AC fans.

By the fall of 1887, Tesla and his sponsors moved to the next phase—filing a patent on the AC motor in order to protect the invention. Peck turned to Parker Page of the respected law firm of Duncan, Curtis & Page. The Harvard-educated Page demonstrated substantial interest in Tesla's motor: his own father had dabbled with electricity and battery-powered locomotives, and his mother had sold an induction-coil patent to Western Union. Page and Tesla worked closely together for several months, deciding to file one application for an entire electricity-transmission system rather than a series of patents for different components. The Patent Office, however, argued that the proposal was too sweeping and demanded separate applications for various apparatuses. In May 1888 Tesla was granted seven patents covering his AC motor and other poly-phase inventions.

Despite Page's expertise and demonstrated success, Tesla could be a testy client. When the Reichsgericht later rejected two of his patent applications, the inventor complained: "I knew when the German patents were sent off that the omissions [of detailed drawing] were very important ones, but at that time you paid no attention to my statements." Tesla criticized: "I want you to understand that I do not exactly blame you, but you must yourself admit that we have two or three times had hard luck, if you want me to use this term, which could have been avoided by careful attention to all formalities."[29] This from the visionary who had disliked drafting; obviously, Tesla had grown to understand the importance of detailed drawings.

Tesla spent the spring and summer of 1888 devising some twenty variations to his poly-phase motor, some to start up under heavy loads, some to run at variable speeds, and others to operate at constant speed. (Those AC motor designs eventually found their niche powering heavy machinery as well as refrigerators, fans, and other appliances.) Page filed patent applications for each, while Peck and Brown developed a business plan to market the devices. Rather than open and manage their own manufacturing plant, which the bureaucracy-hating Tesla would have abhorred, the partners decided to sell the poly-phase patents to other industrialists. Convincing buyers that Tesla's patents offered enormous profit potential meant getting his inventions recognized by the media and scientists. Peck and Brown, in essence, became Tesla's public relations agents.

Unfortunately, the thirty-two-year-old inventor had kept a low profile during his four years in New York. He had not joined any of the professional associations, such as the American Institute of Electrical Engineers, Electrical Club of New York, or National Electric Light Association. Peck and Brown felt Tesla could use the endorsement of a respected expert, so they turned to William Anthony, who had been a professor of physics at Cornell University. To protect against snooping competitors, the inventor and the professor exchanged visits in private, and Anthony came away impressed, claiming Tesla's motors were 50–60 percent more efficient than direct-current models. In a letter to a colleague, Anthony stated, "I [have] seen a system of alternating current motors in N.Y. that promised great things."[30]

With the professor's praise and the patents approved, Peck and Brown arranged for the two leading trade journalists—Charles Price of the *Electrical Review* and Thomas Commerford Martin of *Electrical World*—to visit Tesla's lab. Price quickly wrote a favorable story and Martin became entranced with the inventor, whom he described as having "eyes that recall all the stories one has read of keenness of vision and phenomenal ability to see through things." Finding Tesla to be a "congenial companion," Martin styled their initial conversation as "dealing at first with things near at hand and next . . . rises to the greater questions of life, and duty, and destiny."[31] With his handsome bald head and sculpted mustache, Martin proved to be an effective promoter—flamboyant, aggressive, well-connected, and well-versed in electrical engineering. Known to his friends as Commerford, he had helped to establish the American Institute of Electrical Engineers (the forerunner of today's Institute of Electrical and Electronics Engineers or IEEE).

Tesla and Martin, both born in the same year, complemented each other, with the editor gaining good copy from the brilliant inventor. The author could be gushing and florid, saying "Mr. Tesla has been held a visionary, deceived by the flash of casual shooting stars; but the growing conviction of his professional brethren is that because he saw farther he saw first the low lights flickering on tangible new continents of science."[32] Tesla, in turn, having admired how Edison attracted a steady stream of newspapermen to publicize his Menlo Park feats, relied on Martin for similar assistance. The publicist soon prompted a glowing profile in the *New York World*, Manhattan's widest-circulation daily, with a headline reading "OUR FOREMOST ELECTRICIAN" and a sub-head of "Greater Even Than Edison."[33] A few months later, the *New York Times* promoted a multi-columned profile of Tesla with the subhead: "Advancing with Certainty to Greatest Triumphs."[34]

Still, the reclusive inventor preferred isolation over mingling, and he initially saw little advantage in giving speeches or writing articles. Martin persisted, sensing that Tesla represented a new wave within the increasingly glamorous field of electrical engineering. He wrote: "Tesla stood very much alone, [as] the majority (of the electricians) were entirely unfamiliar with [the motor's] value."[35]

When convinced of the need to garner public attention, Tesla began to eat dinner regularly at Delmonico's, the posh restaurant where he could be seen by the city's wealthy and famous. It was "the one and only time in my life," he said later, when "I tried to roar a little bit like a lion."[36] On occasion—with assistance from Martin, Peck, and Brown—he organized lavish duck dinners for financiers and reporters, followed by tours and demonstrations at his laboratory, which, according to one participant, became "filled with a terrific lightning display, with the snapping, crackling sound, displacing the reverberation of heaven's artillery, and all remarked on the weird and awing effect of the exhibition."[37] Located in the graceful and triangular brownstone at 2 South William Street in New York City, Delmonico's grand dining room featured polished silverware, translucent china, flowing purple curtains, and a bar of rich dark wood. The restaurant introduced the Delmonico steak and lobster Newburg and reportedly launched baked Alaska and eggs Benedict. It attracted the likes of Theodore Roosevelt, Mark Twain, "Diamond Jim" Brady and Lillian Russell, J. P. Morgan, Edward VII, and Napoleon III.

Martin, Peck, and Brown also put Tesla on the lecture circuit, with the highest-profile talk being to the American Institute of Electrical Engineers (AIEE) in May 1888. The inventor, feeling overworked and ill, initially expressed reluctance; only after much prodding did he write out his speech—and then only the night before the talk, in pencil, and hastily.

Speaking in his high-pitched voice, the scientist promised to introduce "a novel system of electric distribution and transmission of power by means of alternate current...which I am confident will at once establish [its] superior adaptability."[38] He offered a mathematical explanation of electromagnetic forces, as well as a basic description of the poly-phase motor. He emphasized that the motor was adaptable and synchronous, meaning it could run at a generator's speed. Although Tesla had discovered revolutionary new principles, his diagrams and explanations were clear and lucid, causing some engineers to believe after the lecture that they had known these approaches all along.

Professor Anthony described his positive tests on two of Tesla's experimental motors, and Thomas Commerford Martin boasted about their efficiency. An awkward silence came over the room when Elihu Thomson, a distinguished inventor, stood to describe his own AC motor that relied on a commutator to create a magnetic repulsive force and turn the rotor. Tesla acknowledged Thomson as "being foremost in his profession," yet he explained why the best AC motor would not employ a commutator. Others agreed Thomson had been bested. The event ended with the AIEE vice president declaring, "I believe that this motor—Mr. Tesla can correct me if I am not right—is the first good alternating current motor that has been put before the public anywhere."[39]

Tesla's poly-phase motor enjoyed several advantages over Thomson's concept. It was cheaper to build since its insulation and windings were simpler, and it did not need costly brushes or commutators. It was less expensive to operate since it required fewer working parts that could wear out quickly. It also was more rugged and could be designed in various sizes, including ones utilizing higher voltages. The U.S. Patent Office confirmed Tesla's primacy and rejected Thomson's patent application, arguing Thomson's "teaser current" approach actually was based on Tesla's invention. (No doubt Thomson subsequently devised improvements for the induction motor, but the legal rejection failed to stop him from claiming to be the real inventor of an AC system. Thomson continued his AC enthusiasm even after his Thomson-Houston Company became part of General Electric in 1892; the following year he helped to build a small hydroelectric facility in Mill Creek, California, which first demonstrated three-phase transmission.)

With the major engineering journals reprinting Tesla's lecture, his poly-phase patent now enjoyed credibility and buzz, allowing Peck and Brown to begin the marketing process. Thomson-Houston had no interest since Elihu Thomson was working on his own AC motor and believed Tesla was a young upstart whose ideas had little value. The best prospect was Pittsburgh-based George Westinghouse, who was making a big bet on alternating-current equipment.

Worried his own company had not been first to develop AC patents, Westinghouse sent a lawyer to Turin, Italy, to secure control of Galileo Ferraris's induction motor, and he purchased rights to the transformer developed by Lucien Gaulard of France and John Dixon Gibbs of England. Intrigued by Tesla's presentation, Westinghouse also sent his legal counselor, H. R. Gardner, to New York to examine the inventor's motors. After a demonstration, Gardner wrote to his boss, "The motors, as far as I can judge from the examination which I was enabled to make, are a success." Admitting Tesla's visions were unique and even revolutionary, Gardner added: "Mr. Tesla struck me as being a straight-forward, enthusiastic sort of a party, but his description was not of a nature which I was enabled, entirely, to comprehend." The lawyer admitted he kept the visit short and tried "to avoid giving the impression that the motor was one which excited my curiosity."[40]

Peck played the role of hard-ball negotiator, hinting that a San Francisco capitalist, a Mr. Butterworth, was willing to put up $200,000 plus a generous royalty of $2.50 for every horsepower sold. Gardner found such terms "monstrous" and he told Peck "there was no possibility of our considering the matter seriously." Peck shot back "that unless [Westinghouse] can let them know by ten o'clock, Friday of this week, whether or not we propose, seriously, looking into the matter, they will accept Butterworth's proposition."[41]

Westinghouse remained interested. Eager for confirmation that Tesla's motors worked as advertised, he dispatched one of his chief scientists, William Stanley, Jr. A "nervous and agile" man of no small talent or arrogance, Stanley arrived at Tesla's workshop and declared the "Westinghouse boys" had already developed an effective AC motor, to which Tesla calmly asked Stanley if he'd like to see his latest model, and Stanley was won over. He admitted, "Their motor is the best thing of the kind I have seen. I believe it more efficient than most DC motors. I also believe [the patent] belongs to them."[42]

Westinghouse desperately wanted to avoid the risk of lawsuits: "If the Tesla patents are broad enough to control the alternating motor business, then the Westinghouse Electric Company cannot afford to have others own the patents."[43] Westinghouse argued the $2.50/horse-power charge "seems rather high, but if it is the only method of operating a motor by the alternating current, and if it is applicable to streetcar work, we can unquestionably easily get from the users of the apparatus whatever tax is put upon it by the inventors."[44]

In mid-July 1888, Westinghouse signed an agreement to own about twenty of Tesla's poly-phase patents "for inventions of converters and alternating and continuous current electric generators, and kindred apparatus," in exchange for 150 shares of his company's capital stock (then valued at some $315,000, or about $7.7 million in today's dollars) as well as the royalty of $2.50 per horsepower for each alternating-current motor sold.[45] Westinghouse, moreover, offered Tesla $2,000 per month (almost $600,000 annually in today's dollars) if he moved to Pittsburgh to help his engineers construct an AC system.[46]

4.

A WHIRLING FIELD
OF FORCE
Pittsburgh

Tesla traveled to western Pennsylvania in late July 1888, hopeful his brilliant work would be appreciated and that he would finally get paid what he deserved. Yet rather than develop the various components of a complete poly-phase system, he and George Westinghouse had to spend substantial time defending their patents from several litigants. Many inventors argued that they'd crossed the finish line before Tesla. Walter Baily, Charles Bradley, and Marcel Deprez, for instance, contended their discoveries anticipated his. Charles Proteus Steinmetz, who would become General Electric's chief scientist, proclaimed he had created the first "monocyclic" AC system, and Elihu Thomson continued to declare his supremacy. Contesting these claims proved to be expensive, and the various and contradictory assertions confused both the public and the courts.

However, in September 1900, after some twelve years of legal wrangling, Tesla won a clear judgment from the U.S. Circuit Court of Connecticut, which glowingly declared: "The search lights shed by defendant's exhibits upon the history of this art only serve to illumine the inventive conception of Tesla.... It was he who first showed how to transform the toy of Arago into an engine of power; the 'laboratory experiment' of Baily into a practically successful motor; the indicator into a driver; he first conceived the idea that the very impediments of reversal in direction, the contradictions of alternations might be transformed into power-producing rotations, a whirling field of force."[1]

In response to those suggesting Tesla simply offered minor modifications to the works of others, Judge Townsend commented on the fundamental nature of Tesla's invention: "The apparent simplicity of a new device often leads an inexperienced person to think that it would have occurred to anyone familiar with the subject, but the decisive answer is that with dozens and perhaps hundreds of others laboring in the same field, it had never occurred to anyone before [Tesla]."[2]

Here at last was vindication. While the judge conceded that the field was packed with inventors, he singled out Tesla as both the pioneer and problem solver who made alternating current work.

Tesla and Westinghouse, despite their different styles, proved to be a better match than Tesla and Edison. Westinghouse complemented Tesla, providing guidance and support. He also offered balance to the unconventional inventor, keeping him grounded in the present so he could envision the future.

Tesla grew to admire his latest partner, a talented inventor in his own right. He found him to be "always smiling, affable, and polite" and would not utter "one word which would have been objectionable, not a gesture which might have offended." Yet, Tesla observed, Westinghouse displayed "tremendous potential energy—even to a superficial observer the latent force was manifest. . . . No fiercer adversary than Westinghouse could have been found when he was crossed."[3]

Both men invented profusely, with Westinghouse holding some four hundred patents, most notably for railroad brakes. Previous devices, known by trainmen as the "arm-strong brake," could barely halt a thirty-mile-per-hour train within 1,600 feet. Westinghouse's initial air brakes cut that distance to five hundred feet, and his refinements—which garnered 103 patents—reduced the length further, avoiding train collisions and curtailing equipment damage.

Westinghouse hoped Tesla's motor could be integrated into his company's single-phase circuits working on a 133-cycle current, since he and his team had neither the money nor interest to develop an entirely new system. Although Tesla preferred his own poly-phase ideal, expecting engineers to adjust to his beautiful configuration, he obediently tried to alter his poly-phase motors. Unfortunately, the alternatives didn't work effectively at 133 cycles. After many frustrating failures, Westinghouse engineers backed down and changed their central-station frequency to suit Tesla's design, trying to find one high enough to keep the lights from flickering but low enough to run the motors efficiently. They settled on 60 cycles and began to manufacture an array of generators and electric motors for trains, pumps, and appliances. Within a decade, the 60-cycle current of the Westinghouse-Tesla partnership became the United States standard for power production.

For a time, Tesla felt happy and productive in Pittsburgh. Guided by a pragmatic boss yet relatively free to invent, he filed some fifteen patent applications in 1889, his most productive patent-generating year. At the age of thirty-three, he also worked with the "Westinghouse boys" to electrify the city's streetcar system and to test different amounts of copper and soft Bessemer steel within his designs, eventually doubling the system's output. The Westinghouse Company, in turn, manufactured almost a thousand Tesla motors that year.

Westinghouse, like Edison, was a decade older than Tesla. His childhood, like Tesla's, was spent with an inventive parent. George Jr. learned his skills at his father's shop in Schenectady, New York. George Westinghouse, Sr., opened that facility in 1856 to build small steam engines, farm machines, and mill works, and the ingenious mechanic eventually acquired several patents for sewing machines, threshers, and winnowers.

Although usually quiet in public, Westinghouse could be engaging and even charismatic. A solid barrel-chested man of six feet, with brawny muttonchops and an abundant mustache, his eyes appeared both intense and genial. He often wore a formal dark-vested suit and sported an ever-present umbrella. In sartorial habits, he and Tesla were well matched. According to a biographer, "With his soft voice, his kind eyes, and his gentle smile, he could charm a bird out of a tree."[4] Tesla described Westinghouse as having "a powerful frame, well proportioned, with every joint in working order, an eye as clear as crystal, a quick and springy step—he presented a rare example of health and strength."[5]

Westinghouse sketched and dictated constantly. At home, he designed on a billiards table, and his limo and private train car served as roving offices. Each morning Westinghouse delivered reams of sketches and directives to his teams, and he demanded fast action. Throughout the afternoons, he dogged the engineers and workers, listening carefully to their concerns and suggestions and using the corner of virtually any table to draw alternate approaches. Tesla observed that the manufacturer loved his work: "Like a lion in a forest, he breathed deep and with delight the smoky air of his factories."[6]

Westinghouse had become a wealthy and successful inventor before he turned his attentions toward electricity. That he survived and thrived in the ruthless world of railroad conglomerates testifies to his business skills and persistence. Railroad barons—including Cornelius Vanderbilt, Jay Gould, and James Hill—forced smaller lines into bankruptcy, consolidated nationwide monopolies, and cold-bloodedly squeezed their suppliers and captive customers. Westinghouse learned of their ruthlessness early when he licensed his first invention—a "car replacer" that moved derailed trains back onto the tracks—to railroad companies, which quickly made slight "improvements" to his device and claimed the resulting patents and profits. When it came to his brakes, however, a wiser Westinghouse refused to sell licenses to the railroads, deciding instead to manufacture the equipment at his own factories in Pittsburgh.

No doubt Westinghouse's electricity work followed Edison's. While the Menlo Park team opened the Pearl Street Station in 1882, Westinghouse's first major lighting projects—the Windsor Hotel in New York and the Monongahela Hotel in Pittsburgh—started operations in 1886. Later that year, he installed in Trenton, New Jersey, his first "centralized" power station, composed of six 100-volt DC dynamos, each of which could power three hundred lamps. Yet Westinghouse expanded rapidly. Within three years, his company had connected enough generators to illuminate more than 350,000 incandescent bulbs.

Westinghouse built his electric company largely by acquiring other people's patents rather than relying on his own discoveries. Willing to try innovative technologies, the bold and ambitious businessman purchased well, buying the floundering United States Electric Light Company, which owned the important lamp patents of Hiram Maxim and William Sawyer. Also demonstrating an entrepreneur's knowledge of the law, Westinghouse filed suit against the Edison Electric Light Company for infringing on Sawyer's patent. Edison, of course, fought back, but he remained stubbornly committed only to his own inventions, while Westinghouse courted new ideas.

Tesla was initially content to be in Pittsburgh, yet Westinghouse engineers grew frustrated with him. Some expressed envy of the generous deal he received for an invention they thought they had developed. Others simply could not tolerate what they considered to be a pompous foreigner.

And it wasn't long before the inventor became bored with practical engineering and the bureaucracy of running a business. This was a man who had visions of changing the world. He loathed the mundane tasks associated with tests and adjustments, and he disdained others second guessing his insights. He increasingly felt trapped in the present. In the early summer of 1889, when Westinghouse engineers used graphite bearings despite Tesla's insistence that they would overheat, Tesla abruptly left Pittsburgh for Paris. There he toured the new Eiffel Tower, attended the Congrès internationale des électriciens, and heard lectures from some of the world's leading scientists, including Vilhelm Bjerknes, a Norwegian physicist who had worked with Heinrich Hertz on electromagnetic waves.

The short French outing offered to Tesla an overview of other alternating-current experiments that "opened up wonderful possibilities if producible in practical ways; there were the currents of many hundreds of thousands of amperes, which appealed to the imagination by their astonishing effects; and most interesting and inviting of all, there were the powerful electrical vibrations with their mysterious actions at a distance."[7] Largely because he learned European inventors were tackling high voltages, he decided to concentrate on high-frequency phenomenon, sensing that a larger number of cycles per second could make lamps glow brighter, transmit electricity more efficiently, and send power as well as communications wirelessly.

George Westinghouse tried to persuade him to return to Pittsburgh, offering an even more generous salary and a well-equipped laboratory. No doubt Tesla would have become a very rich man if he worked within such a large industrial concern, but he favored freedom. That decision proved to be a turning point for Tesla, particularly in terms of productivity.

Tesla, who needed equilibrium between the worlds of idealism and reality and between the present and the future, felt in Pittsburgh that his balance had tilted too much toward "this world." And when he sensed being limited by the mundane status quo, Tesla became unhappy and uncreative.

"I was not free at Pittsburgh," he declared. "I was dependent and could not work." Seeking to be unburdened from routine tinkering, he longed for "inventing methods and devising means for enabling scientific men to push investigations far out into these practically unknown regions."[8] As a result of leaving, he later claimed, "Ideas and inventions rushed through my brain like Niagara."[9]

5.

AS REVOLUTIONARY AS GUNPOWDER WAS TO WARFARE

New York City

When Tesla returned to New York City in mid-summer 1889, he asked Peck and Brown to locate a slightly larger laboratory space, and with initial revenue from his patent sales to Westinghouse he moved into the Astor House, which had been the city's first luxury hotel. The scientist, moreover, hired a few specialists to help in the lab—including a glassblower, two mechanics, and an arc-lighting expert—although he relied mostly upon Szigeti, "a man," wrote Tesla, "who had a considerable amount of ingenuity and intelligence. . . . He was not exactly a theoretical man, as myself, but he could understand every idea fully."[1]

Tesla each morning walked the sixteen blocks from his hotel, located on the corner of Broadway and Vesey Street, to his lab within a five-story factory building on the corner of Grand and Lafayette. Again demonstrating his living between two worlds, Tesla dressed for the opera in his regular frock coat and gloves, but he passed an array of sweatshops filled with seamstresses and woodworkers. He usually arrived around 10:00 a.m., sharing the streets with double-parked horse-drawn wagons and vendors hawking clothing, buttons, and glassware.

The inventor spent most of his time at the lab either alone or with one or two assistants; they often worked through the night so as not to be distracted. Other than periodic trips to Pittsburgh to check on the manufacturing of his motors, he enjoyed the isolation, motivated by the belief he was on the cutting edge of a new age of discovery.

Tesla initially tried to obtain high frequencies by adapting rotary generators to operate at substantial speeds. At about twenty thousand cycles per second, however, those machines began to break down. Since rapid reversals of the current caused damaging heat, Tesla had to strike a balance between speed and temperature, what he described as "a thoroughly Wagnerian opera" in which he struggled "to get from the Scylla to the Charybdis."[2]

To deal with the heat, Tesla eventually replaced the melting insulation with an air gap, and he made sure the hot iron core couldn't move to different positions. He then cleverly adjusted an induction coil—which produces high-voltage pulses from low-voltage current—in order to obtain resonance, in which one portion of the circuit drives another to oscillate with greater amplitude, allowing Tesla to produce a current alternating at thirty thousand times per second. Although he referred to this invention as an oscillating transformer, calling it "as revolutionary as gunpowder was to warfare,"[3] the device quickly became known as the Tesla coil.

That high-frequency device provided for the first time a smooth, continuous electrical current at thousands of volts and at any specific frequency. It opened the door to advances in electrical lighting, phosphorescence, X-ray generation, electrotherapy, atom splitting, and the transmission of electrical energy without wires. Until the 1920s and the advent of vacuum tube oscillators, that groundbreaking machine also formed the core of radio transmitters for wireless telegraphy.

Tesla felt his coil would make electrical lighting more efficient. Edison's incandescent bulb, essentially a burning filament within a vacuum, heated up and resulted in a huge waste of energy that Tesla was determined to avoid. He designed various gas-filled cylinders that shone when energized by his new oscillating transformer, and he devised others with phosphorescent coatings that glowed when excited by the coil's thrusts. Tesla was doing the impossible—moving power across a room with neither wires nor groundings. These wireless lamps, according to one reporter, were "devices of mystic origin."[4] (The glowing tubes, which Tesla called "flaming swords," looked like the light-sabers imagined almost a hundred years later by George Lukas in *Star Wars*.)

The inventor initially kept his wireless lighting thoughts and efforts to himself; one early morning he sent his lab assistants out to get some food and when they returned Tesla stood in the middle of the lab holding two glowing long glass tubes that were not connected by a wire to his high-frequency coil. "I waved them in circles round and round my head," the enthusiastic Tesla explained, as he would in later demonstrations at Columbia College. "My men were actually scared, so new and wonderful was the spectacle.... They thought I was some kind of a magician or hypnotizer."[5]

Tesla appreciated the popular lure of such new wonders. He enjoyed the way spectators viewed his glowing tubes "with amazement almost impossible to describe," yet he worried that fickle crowds "soon become indifferent to them. The wonders of yesterday become today's common occurrences." Commenting on other barriers to innovation, he complained that people, particularly investors, resisted radical ideas, even ideas that improved the status quo. He also grumbled that entrenched interests opposed change, saying, "Perhaps the greatest impediment (to invention) is encountered in the prejudicial opinions created in the minds of experts by organized opposition."[6]

Despite such obstacles, Tesla developed more than fifty versions of his coil, some with windings that were cylindrical, conical, or flat, and some insulated by oil or air. He also devised spark gap transmitters (that generated electromagnetic waves), oil-insulated transformers (that raised or lowered voltages), and condensers (now known as capacitors, which store electricity temporarily). His goal was simplicity, "doing away with all packings, valves and lubrication" and producing an "absolutely steady and uniform" current.[7] By the start of the twentieth century, the Tesla coil was a standard part of virtually every college science laboratory in the United States and Europe. (Almost thirty years later, scientists at the Carnegie Institution of Washington used a 5-million-volt Tesla coil in their pioneering attempt to smash the atom.)

The experiments with Tesla's coil were not without their dangers. One evening, the inventor, even after taking his normal precautions, claimed to have been "almost killed today" after receiving a 3.5-million-volt shock from one of his machines. "The spark jumped three feet through the air," Tesla revealed, "and struck me here on the right shoulder. I tell you it made me feel dizzy. If my assistant had not turned off the current instantly it might have been the end of me. As it was, I have to show for it a queer mark on my right breast where the current struck in and a burned heel in one of my socks where it left my body."[8]

The combination of Tesla's AC motor and high-frequency coil catapulted him into the front ranks of inventors, placing him on a par in public popularity and among his scientific colleagues with Thomas Edison and Alexander Graham Bell. He was thirty-three years old and full of ideas and promise.

Three unrelated events, however, tempered Tesla's progress and outlook. The first involved money, particularly the royalty provided by George Westinghouse. Tesla's inventions had helped Westinghouse expand rapidly—from $800,000 in sales in 1887 to $4.7 million in 1890. To build factories and hire engineers, the Pittsburgh businessman had borrowed heavily, and when the national economy tumbled in November 1890, Westinghouse's bankers called in their loans, and he was forced to turn to Wall Street investors who demanded a financial reorganization that would rein in the firm's spending.

J. P. Morgan and other rich financiers, in fact, sought to starve Westinghouse out. In that era of robber barons and giant monopolies, they thought they could do to power generation what they had done to steel, oil, and sugar production. They essentially wanted Westinghouse to sell his company at a cheap price and allow them to create a conglomerate that would control the emerging electricity industry.

Westinghouse balked but had few options. His best hope was to ask Tesla to renegotiate their contract for a $2.50-per-horsepower royalty for each motor sold, so he approached the inventor directly and declared, "Your decision determines the fate of the Westinghouse Company." According to Tesla's first biographer, who tended to glorify and dramatize the scientist's actions and words, Tesla responded, "If I give up the contract, you will save your company and retain control? You will proceed with your plans to give my polyphase system to the world?"

Although the episode is not mentioned in Westinghouse's official biography, the businessman supposedly replied, "I believe your polyphase system is the greatest discovery in the field of electricity. . . . It was my efforts to make it available to the world that brought on the present difficulty. But I intend to continue, no matter what happens, with my original plans to put the country on an alternating-current basis."

"Mr. Westinghouse," Tesla responded theatrically, "you have been my friend; you believed in me when others had no faith; you were brave enough to go ahead . . . when others lacked courage; you supported me when even your own engineers lacked vision to see the big things ahead that you and I saw. . . . You will save your company so that you can develop my inventions. Here is your contract and here is my contract—I will tear both of them to pieces, and you will no longer have any troubles from my royalties. Is that sufficient?"[9]

It's hard to imagine how anyone would walk away from that much money. No doubt the inventor was a poor bargainer and businessman; in this instance, he actually had the negotiating advantage since he held the underlying patents and a signed contract. Tesla also was a romantic, believing he needed to sacrifice in order for the world to enjoy his inventions; yet despite such selfless claims he clearly enjoyed the fancy meals and lodgings money could provide. No doubt he truly respected Westinghouse, was grateful for the businessman's early support, and believed he was "the only man on the globe who could take my alternating current system under the circumstances then existing and win the battle against prejudice and money power."[10] Tesla, moreover, remained extremely confident in his own ability to invent greater devices. Still, the decision to tear up the Westinghouse contract changed Tesla's life in ways he couldn't have imagined, forcing him to face financial challenges for the rest of his life.

The contract termination helped Westinghouse attract new financing and maintain limited control of his company. Yet the loss of royalties meant Tesla sacrificed substantial sums—probably hundreds of millions in today's dollars—that might have allowed him to finance better laboratories and expensive experiments. Tesla, however, maintained his personal admiration of Westinghouse, and the two would work together to power the World's Columbian Exposition in Chicago and to tap Niagara Falls' hydroelectric potential.

Given that Tesla's invention hugely benefitted the Westinghouse Electric & Manufacturing Company and that the chairman greatly admired the inventor, why didn't George Westinghouse reward Tesla when the company became prosperous? The answer seems to be that Wall Street bankers and investors, who increasingly controlled Westinghouse's finances, paid only for understandings bound by legality rather than loyalty.

The second event that inhibited Tesla was the unexpected death of Charles Peck, a key supporter, promoter, and advisor. The lawyer became sick, moved to Asheville, North Carolina, and died there in the summer of 1890. As evidenced by his successful promotion of Tesla's AC-motor patent and his subsequent negotiations with Westinghouse, Peck had provided the business acumen and reality check the idealistic inventor so desperately needed.

And the third event that affected Tesla was deeply personal. Tesla had few equals; however, Anthony Szigeti had been his "best friend" and confidante for nearly a decade. Szigeti was with Tesla in a Budapest park when he outlined his alternating-current motor in the sand with a stick. The two of them worked side by side in various laboratories and often shared long walks and lengthy dinners. Their relationship had become more than one of employer to employee. The inventor referred to Szigeti as "the only one who had supported me during my first attempts and whom I loved dearly because of his virtues and respect."[11] In early 1890, Szigeti decided to independently develop his own inventions, particularly an innovative compass for ships. When the two friends next met, after five months of separation, Tesla, who felt abandoned, scoffed at Szigeti's efforts, saying the compass already had been developed by Sir William Thomson (later to be known as Lord Kelvin), which prompted the proud Szigeti to leave for good, initially to South America. In late 1890, Anthony Szigeti died without warning. Tesla's only initial reaction to what must have been a severe blow was in a note to his family: "I am completely alienated and sometimes it is difficult."[12] Yet twenty years later, Tesla expressed intense pain associated with the loss: "I would have much desired to see him, because I would have wanted him."[13]

(Some biographers have pointed to such comments and claimed Tesla was homosexual. No doubt he enjoyed close relations with a few men throughout this life, yet there's no definite evidence to prove or disprove the assertion. In fact, Tesla's germophobic refusal to touch others makes it hard to believe he enjoyed romance with either men or women.)

Tesla buried his frustrations and anguish in work. In addition to increasing the frequencies from his oscillating transformer, he developed a spherical carbon button that glowed when placed at the end of a wire connected to one terminal of his coil. Compared to Edison's incandescent bulb, this button lamp provided twenty times as much light for the same amount of current consumed. It featured refractory material within a closed and airless globe being bombarded by molecules activated by his rapidly alternating current. As Tesla put it, the molecules struck the "carbon button" many times a second "and at enormous speeds, and in this way, with trillions of invisible hammers, pound it until it gets incandescent."[14] This high-intensity button lamp— a directed ray, what Tesla described as a "pencil thin" line of light, that proved to be the precursor to a laser beam—also could "vaporize" matter and melt even zirconia and diamonds, the hardest known substances. In what he called "a striking experiment," Tesla zapped tinfoil, causing it to "not only melt, but it would be evaporated and the whole process took place in so small an interval of time that it was like a cannon shot."[15] The inventor predicted his button lamp could heat and mold hard metals and thereby would revolutionize the field of metallurgy.

◎ ◎ ◎

Tesla increasingly got dragged into the "War of the Currents," the high-stakes battle between the advocates of direct and alternating current. In the late 1880s, the outcome was not clear. Edison and his financial backers, including the powerful J. P. Morgan, held the advantage, certainly the most projects and the most funding. Tesla and George Westinghouse had what they thought was a more powerful and efficient system. Both electrical approaches, however, competed against natural gas companies that were not willing to cede their dominance over the markets for lighting and heating. The stakes, and the potential profits, were substantial.

Westinghouse had initially taken the high road in this "war" by inviting Edison in 1888 to tour his facilities in Pittsburgh, writing him a personal note: "I believe there has been a systematic attempt on the part of some people to do a great deal of mischief and create as great a difference as possible between the Edison Company and The Westinghouse Electric Co. when there ought to be an entirely different condition of affairs."[16] Edison rebuffed Westinghouse's offer, claiming "my laboratory work consumes the whole of my time."[17] And he doubled down on the distinction between AC and DC, saying "direct current was like a river flowing peacefully to sea, while alternating current was like a torrent rushing violently over a precipice."[18]

Edison launched a public relations campaign in fall 1890 against alternating current, including an eighty-four-page diatribe that assailed Westinghouse and pleaded incongruously with "all electricians who believe in the future of electricity . . . to unite in a war of extermination against cheapness in applied electricity, whenever they see that it involves inefficiency and danger."[19] Westinghouse shot back with his own brochure challenging Edison's own safety record: "Of the 125 central stations of the leading direct current company [Edison's] there were numerous cases of fire, in three of which cases the central station itself was entirely destroyed."[20]

The war then turned gruesome. Edison dispatched an associate— Harold Brown, who Edison claimed was not an official employee but who had access to the Menlo Park laboratory and regularly shared conspiratorial letters with the famous inventor—to orchestrate a public display in which he placed electric wires on the head of a sixty-seven-pound dog described as vicious. Direct current, even at varying voltages, did little to the black retriever, but high-voltage alternating current killed it immediately. According to one journalist, "Many of the spectators left the room unable to endure the revolting exhibition."[21] Not stopping with a single dog, Brown conducted the same "experiment" on fifty different creatures, including cats, calves, and a horse, in order to demonstrate AC was the perfect medium for electrocution.

Thomas Edison actually suggested to the New York legislature that electrocution by alternating current offered a more humane method of capital punishment than hanging. The quickest and most painless death, he wrote, "can be accomplished by the use of electricity, and the most suitable apparatus for the purpose is that class of dynamo-electric machine which employs intermittent currents. The most effective of those are known as 'alternating machines' manufactured principally in this country by Geo. Westinghouse." Edison went so far as to recommend that the new capital-punishment procedure be named "Westinghoused," and he published promotional brochures warning homeowners: "Don't let your house get Westinghoused."[22]

Edison himself outlined specific electrocution procedures, proposing "to manacle the wrists with chain connections, place . . . the culprit's head in a jar of water diluted with caustic potash and connect therein . . . to a thousand volts of alternating current."[23] His lobbyists convinced New York State to power these electric chairs at the Auburn and Sing Sing State Prisons with secondhand Westinghouse generators (Westinghouse refused to sell any). The first human electrocution took place on August 6, 1890, of William Kemmler, a thirty-year-old alcoholic who had murdered his common-law wife with a hatchet, smashing her skull twenty-six times. Although guilty of a heinous crime, Kemmler suffered a death that was described as "an awful spectacle, far worse than hanging," with witnesses saying the criminal's spinal cord exploded into flames as the current ran through his body for seventeen seconds. Furthermore, "to the horror of all present, the [prisoner's] chest began to heave, foam issued from the mouth, and the man gave every evidence of reviving."[24] A second jolt finally caused Kemmler to go rigid, but his clothes and skin caught fire and the *New York Times* reported, "The stench was unbearable."[25]

The execution turned public opinion against Brown and Edison. One disgusted eyewitness declared, "I would rather see ten hangings than one such execution as this." Other newspapers likened the work to barbarians and torturers within "the darkest chambers of the Inquisition of the 16th Century."[26] Edison embarrassed himself further by suggesting that future electrocutions avoid placing the shock on the criminal's head, since hair was nonconductive; instead, he argued, "The better way is to place the hands in a jar of water . . . and to let the current be turned on there."[27]

Tesla entered the debate by writing about his inventions. In the February 1891 issue of *Electrical Engineer,* he described how alternating current could send power safely over long distances. Edison responded bitterly, and the two inventors sparred in a series of trade-journal letters throughout the winter and spring. Raising the stakes, the scientist decided to lecture to the American Institute of Electrical Engineers, this time on May 20, 1891, at Columbia College in New York City.

As mentioned at the beginning of this book, Tesla, with a thick Eastern European accent but impeccable English grammar, started this lecture and show by expressing wonder at the forces of nature: "What is electricity," he asked provocatively, "and is there such a thing as electricity?" He suggested the laws of electricity change when power moves from direct to alternating currents. "When the currents are rapidly varying in strength," he observed, "quite different phenomena, often unexpected, present themselves, and quite different laws hold good." He discussed electromagnetic waves, electrostatic thrusts, and the muscle of high-frequency AC. One reporter described the scientist's insights as "epoch making."[28]

Yet Tesla's demonstrations were what captivated the crowd of experimenters, reporters, and financiers. For a man claiming to be shy, he became a master showman. As he increased the current's frequency, he illuminated lamps and gas-filled tubes without wires, heat, or flames whenever he passed them between the electrostatic field created by charged zinc plates on either side of the stage. Debunking Edison's claims about AC's dangers, the "magician" also sent thousands of volts of alternating current through his body, which caused his coat to radiate a soft blue and his fingers to release tiny sparkles, yet the inventor remained very much alive.

The glowing tubes were a crowd favorite, with one reporter writing "they appeared like a luminous sword in the hand of an archangel representing justice."[29] The inventor's ideal, in fact, was to create a powerful, rapidly alternating electrostatic field that would allow "an illuminating device" to be lighted anywhere in a room "without being electrically connected to anything." Yet the dreamer admitted: "How far this principle is capable of practical application, the future will tell."[30]

These wireless transmissions inspired Tesla, who described them as "the first evidence that I was conveying energy to a distance, and it was a tremendous spur to my imagination."[31] He had separated the energized plates for the Columbia College lecture by fifteen feet but believed wireless electromagnetic charges could travel much farther.

Tesla concluded his two-hour lecture with another tribute to electricity's wonder. "The field is wide and completely unexplored," he said, "and at every step a new truth is gleaned, a novel fact observed." The concept of "availing ourselves of [nature's] energy more directly" gave him hope "humanity will advance with great strides." Ever the optimist, he concluded, "The mere contemplation of these magnificent possibilities expands our minds, strengthens our hopes, and fills our hearts with supreme delight."[32]

Numerous other showmen were demonstrating scientific wonders and attracting enthusiastic crowds during this period. Con men abounded, such as Walter Honenau claiming to obtain free energy from a "hydro-atomizer," Gaston Bulmar selling special pills that supposedly turned water into gasoline, and Victor "the Count" Lustig introducing a machine that purportedly converted white paper into crisp twenty-dollar bills. Perhaps the most noteworthy snake-oil salesman was John Ernst Worrell Keely, a former "circus slight-of-hand performer" who claimed to have developed a perpetual motion motor. With his scientific sounding explanations—such as "the reflex action of gravity" or "depolar etheric waves"—Keely obtained regular coverage from the New York dailies and an investment from none other than John Jacob Astor. Even luminaries like Thomas Edison advanced dubious but popular assertions, such as his professed ability to photograph thoughts.

No doubt Tesla, too, made outlandish claims, such as his prognostication that he would communicate with other planets. Yet even though most of his visions were realized, Tesla was forced to compete with bogus inventors for the public's attention and the investment community's resources.

Still, Tesla's reputation continued to grow and reports of his successes made headlines in both the United States and Europe, where he became something of a hero to his family and to Serbs in general. A brother-in-law wrote, "We think about you even in [our] dreams."[33] Tesla began sending money to his mother and sisters, although his letters back home were addressed to their husbands and he noted, "Somehow it is hard to correspond with the ladies."[34] His sisters begged regularly for some correspondence, complaining "We all feel badly that you did not write,"[35] or "a simple word from you with your handwriting would stop a million tears and strengthen me for future struggles."[36] One sibling even pleaded, "We do not have anyone except you."[37]

Tesla also received a string of requests for money or investments from uncles, nephews, cousins, and even great nephews—several of whom he did not know. One cousin, who introduced herself as the youngest daughter of his Aunt Smiljana, begged: "We are now in the greatest need and by the end of the year [our bills] must be paid."[38] Others sought jobs in Tesla's laboratory for themselves or relatives.

Tesla's family kept up with European newspaper accounts of his accomplishments, and they responded with commendations as well as concerns about the overworked inventor who, sister Marica said, maintained a "frail body construction."[39] Tesla periodically would provide his own updates as well as boasts: "It is difficult to give you an idea how I am respected here in the scientific community."[40]

Although busy with his research and presentations, Tesla completed his citizenship requirements in late July 1891. He expressed great happiness in being an American, saying his official papers, which he kept in a safe throughout his life, were more important than any of his "diplomas, degrees, gold medals, and other distinctions [that] are packed away in old trunks."[41] When asked a few years later if he were a good American, he declared, "I was a good American before I ever saw this country . . . What opportunities this country offers a man! Its people are a thousand years ahead of the people of any other nation of the world."[42]

Tesla, however, never abandoned his pride in being born and raised a Serb. His most valued title, "Grand Officer of the Order of St. Sava," was bestowed in Belgrade, the Serbian capital, by King Alexander I Obrenovic. Remembering his ancestors' struggles against the "unspeakable Turk," he reflected on how Serbs developed notable qualities of bravery and sagacity while maintaining a sense of patriotism and independence. "If [my] hopes [of bringing some of my ideas to the benefit of all humanity] become one day a reality," he said, "my greatest joy would spring from the fact that this work would be the work of a Serb."[43]

6.

ORDER OF THE
FLAMING SWORD

Europe

A s Tesla's fame grew, so did the number of his legitimate competitors. His presentations and articles certainly spurred others to play off his ideas, yet numerous inventors were independently pursuing their own approaches to a practical motor and power distribution. British journals continued to credit Galileo Ferraris with developing the first rotating-field motor. F. A. Haselwander, a German engineer, claimed to have invented the initial poly-phase machine. Oskar von Miller, another German scientist, insisted that he offered the first demonstration of long-distance electricity transmission.

Tesla in 1892, at the age of thirty-six, decided to travel to Europe to defend his own patent claims, as well as to market his inventions to manufacturers in Britain and on the continent. Unlike his first trip across the Atlantic eight years earlier, Tesla returned in a first-class cabin and with far more than four cents in his pockets, thanks to the continued, although diminished, payments from Westinghouse. Building on the success of his Columbia College lecture, he secured speaking invitations in both London and Paris.

On the back end of his trip, he planned to visit his mother in Gospic. They had not been together for almost a decade, although they maintained fairly regular correspondence. He continued to refer to her as his inventing inspiration, and he professed "a consuming desire to see her again."[1]

The Atlantic Ocean crossing gave Tesla a chance to reflect on his achievements and motivations. His New York City demonstration and the introduction of a successful motor had propelled him to become the toast of the town, yet he felt isolated after the deaths of his best friend and his valued business advisor. What would inspire him going forward? Although he enjoyed expensive dinners at Delmonico's, his tearing up of the Westinghouse royalty contract demonstrated that money was not his prime motivation. No doubt he worked hard to establish and maintain his image as a great inventor, yet he possessed a certain modesty. What Tesla recognized on the ship's deck was that his real love was inventing, to pursue the uncertain but glorious process of trying to realize his visions.

The inventor was invited to address the Institution of Electrical Engineers at the Royal Institution in London on February 3, 1892. Some eight hundred people in evening dress filled the institution's amphitheater. The audience included the leaders of England's scientific community, and Tesla began by praising one of their own, Sir William Crookes, the renowned chemist who had conducted early experiments with radiation. Tesla revealed that his reading of a "fascinating little book" by Crookes many years before had prompted his interest in electricity.

Tesla spoke humbly at first. "The results which I have the honor to present before such a gathering I cannot call my own. There are among you not a few who can lay better claim than myself to any feature of merit which this work can contain." One reporter who agreed wrote: "Seldom has there been such a gathering of all the foremost electrical authorities of the day, on the tiptoe of expectation."[2]

As in previous lectures, the scientist admitted astonishment at nature's power. With enthusiasm and delight, he asked, "Can there be a more interesting study than that of alternating current?" Electricity, he declared, "takes the many different forms of heat, light, mechanical energy, and ... even chemical affinity. ... All of these observations fascinate us."[3] He claimed his goal was "to transform, to transmit, and direct energy at will." For the passionate Tesla, "the mass of iron and wires [associated with electricity generation and distribution] behaved as though . . . endowed with life."[4]

Despite such awe, Tesla admitted the basics of electricity generation were surprisingly ordinary: "We wind a simple ring or iron with coils; we establish the connections to the generator, and with wonder and delight we note the effects of strange forces which we bring into play, which allow us to transform, to transmit, and direct energy at will." He argued, however, that his newfound ability to alternate the current to very high frequencies produced astonishing results, which were just beginning to be revealed and appreciated.[5]

Tesla treated the two-hour lecture as a seminar for other inventors, providing advice to aid their experiments. Based on his own trials, for instance, he suggested researchers coat the bases of their lamps with aluminum "on account of its many remarkable properties." He also offered them motivation. "Each day we go to our work," he said, "in the hope that someone, no matter who, may find a solution [to] one of the pending problems—and each succeeding day we return to our task with renewed ardor."[6] Countering his previous claims that he designed machines only in his head, Tesla admitted testing an array of materials for the wire to his carbon button: "I have tried at first silk and cotton covered wires with oil immersions, but I have been gradually led to use gutta-percha covered wires, which proved most satisfactory."[7]

The showman began the evening's entertainment portion, as he had at Columbia College, by walking around the stage between two electrified zinc sheets and holding a long glass tube that glowed "with a brilliant lambent flame from end to end." Standing on an insulated platform, he directed electrical "streams upon small surfaces" and produced "magnificent colors of phosphorescent light" simply by picking up tubes. Tesla, according to a reporter, "recalled to everyone the idea of the magician's enchanted wand."[8]

The inventor kept offering "wonder after wonder." Six-inch sparks leapt between two balls, foot-long wires glowed blue, and bulbs radiated within the electrostatic field. He showed how his motor could run on a single wire, a novel feat in itself, but he brashly suggested wires may not be needed at all since power could be drawn from the earth or atmosphere. He honored Lord Kelvin, the renowned physicist who was still living but not among the onlookers. Tesla spelled out his common name, William Thomson, in light, written with illuminated glass tubing. The crowd applauded with each demonstration, and an English commentator noted: "The interest of the audience deepened into enthusiasm."[9]

For the lecture's conclusion, reported *Electrical Review*, "Mr. Tesla tantalizingly informed his listeners that he had shown them but one-third of what he was prepared to do, and the whole audience ... remained in their seats, unwilling to disperse, insisting upon more."[10] His goal, he claimed, had been to "point out phenomena or features of novelty" in order to "advance ideas which I am hopeful will serve as starting points of new departures."[11] During the question-and-answer period, someone asked if the scientist felt pain as electricity flashed through him, and Tesla responded, "A spark, of course, passes through my hands, and may puncture the skin, and sometimes I receive an occasional burn, but that is all; and even this can be avoided if I hold a conductor of suitable size in my hand and then take hold of the current."[12]

Lord Rayleigh, who had isolated argon and would in 1904 receive the Nobel Prize for physics, closed out the event by commending the inventor for having "the genius of the discoverer."[13] Rayleigh's memorable praise motivated Tesla, who recalled, "[Rayleigh] said that I possessed a particular gift of discovery and that I should concentrate upon one big idea."[14] Although he still had to win the "War of the Currents," Tesla pledged to focus on the wireless transmission of messages and power, the ideal that would obsess him for the rest of his life.

Most attendees, according to *Electrical Engineer*, were "spellbound" by Tesla's "easy confidence and the most modest manner possible displaying his experiments, and suggesting, one after another, outlooks for the practical application of his researches."[15] That magazine ran a glowing profile and declared Tesla's research on electrical motors and coils preceded Ferraris and Haselwander.[16] A separate trade journal credited Tesla with giving many of those attending "for the first time apparently limitless possibilities in the applications and control of electricity."[17] Yet at least one participant wondered if the presentation was more showmanship than science; *The Electrician*'s editor complained that Tesla "did not write and read a paper, nor did he give a lecture, and he was so occupied in waving long glowing electrodeless tubes in the air, and lighting up of ordinary incandescent lamps by a current taken through his body, that he had no time to explore 'how it was done.' Nor, I think, could he."[18]

After Tesla's Royal Institution debut, British notables lined up to meet the new sensation. Imagine the thrill for a young immigrant, whose own father never acknowledged his talents or accomplishments, to suddenly be venerated by the science nobility. Still, Tesla did his best to avoid shaking hands with or getting too close to his admirers.

Ambrose Fleming, inventor of the vacuum tube, invited Tesla to visit his laboratory at University College and declared that "no one can doubt your qualifications as a magician of the first order." The English aristocrat went so far as to dub the American inventor a charter member of his newly created "Order of the Flaming Sword."[19]

James Dewar, the Fullerian Professor of Chemistry at the Royal Institution, begged Tesla to give a second lecture, and he proved to be very persuasive. According to Tesla, the Scotsman "pushed me into a chair and poured out half a glass of a wonderful brown fluid which sparkled in all sorts of iridescent colors and tasted like nectar. 'Now,' said he, 'you are sitting in Faraday's chair and you are enjoying whiskey he used to drink.'"[20] The honor of being treated like Michael Faraday, the brilliant scientist who in the 1830s discovered electromagnetism, overwhelmed the thirty-six-year-old Tesla, who agreed to another two-hour talk and demonstration.

A few days later, a weary Tesla traveled to Paris to deliver a third lecture, this time before the Société française de physique and the Société internationale des électriciens. The City of Light had become even more illuminated in the three years since he had attended the electrical congress, and Tesla himself had become something of a luminary. While he must have been exhausted, he nonetheless won over the crowd with his showmanship, revolutionary insights, and wild predictions. One attending electrician noted, "The young scientist is ... almost as a prophet. He introduces so much warmth and sincerity into his explanations and experiments that faith wins us, and despite ourselves, we believe that we are witnesses of the dawn of a nearby revolution in the present processes of illumination."[21]

Reporters also commented on the scientist's mesmerizing qualities. "Tesla's bright eyes glowed as he spoke of his work," wrote one journalist. "Leaning forward, peering almost each moment into the eyes of his [audience] to make sure that his meaning has been understood, he proved a talker from whose train of reasoning there was no escape while a man was under his influence."[22] According to *Electrical Review*, "No man in our age has achieved such a universal scientific reputation in a single stride as this gifted young electrical engineer."[23]

Still, Tesla attracted critics. William Stanley, the American physicist who worked for George Westinghouse, argued that he was the true inventor of an alternating-current system, and Thomas Edison continued to assert the advantages of direct current. A few simply doubted Tesla's claims, with one magazine proclaiming, "It is questionable whether the Tesla motor can ever be a success."[24]

A persistent Tesla capitalized on his newfound fame to promote his foreign patents to a variety of French manufacturers. This unceasing outreach had already begun to deplete the scientist when he received a telegram from his Uncle Petar that his mother was seriously ill. "[I] was just coming to from one of my peculiar sleeping spells, which had been caused by prolonged exertion of the brain," Tesla remembered, in a fashion similar to the prophetic and disturbing visions he encountered in his youth. "Imagine the pain and distress I felt when it flashed upon my mind that a dispatch was handed to me at that very moment bearing the sad news that my mother was dying."[25]

Nikola rushed to Gospic, where his three sisters embraced him. He walked quickly from the train station through town, which had changed little except for the addition of electric street lamps. When he reached his mother's side, she whispered, "You've arrived, Nidzo, my pride."[26]

Tesla cried at the sight of his beloved inspiration, now ashen and frail. The two had not spoken in about ten years, yet they said little. The son simply maintained a close watch that day and evening by Djuka's bed and witnessed her "agony."

Since Tesla was "completely exhausted by pain and long vigilance,"[27] a family member eventually relieved him from the vigil and another took him to an apartment two blocks away in order to sleep uninterrupted. In the midst of his slumber, in what he later described as a supernatural experience, Tesla "saw a cloud carrying angelic figures of marvelous beauty, one of whom gazed upon me lovingly and gradually assumed the features of my mother. The vision slowly floated across the room and vanished, and I was awakened by an indescribably sweet song of many voices. In that instant a certainty, which no words can express, came upon me that my mother had just died. And it was true."[28]

Djuka was sixty years old when she passed away at three o'clock in the morning of April 4, 1892. Because she descended from three generations of religious leaders, six Serbian Orthodox priests officiated at her funeral, and she was buried next to her husband. Their white obelisk headstones are in the Jasikovac cemetery in Divoselo.

Devastated, Tesla remained in Gospic for another six weeks. "I don't have to tell you that I am very sad and holding myself in restraint," he wrote to a relative. "I was afraid of this event a while ago, but the blow was heavy."[29] Even thirty years later, Tesla declared, "The mother's loss grips one's head more powerfully than any other sad experience in life."[30]

Tesla's dream of his mother dying haunted him. While visions fed his inventing, for some reason he needed to explain away this "strange manifestation." In fact, he never clarified the line between images that prompted his creativity and apparitions that he considered weird and demanded clarification. Numerous contemporary and respected researchers would have accepted a paranormal explanation for the dream about Djuka's death.[31] Sir William Crookes, known for his systematic research on cathode rays, also experimented with mental telepathy, human levitation, and photographs from séances, and he helped found the Society of Psychical Research with Lord Rayleigh, Oliver Lodge, and other notables. Even Thomas Edison studied telepathy and developed a "telephone" to help spiritualists communicate with the deceased. Tesla, however, sought a more rational account, and he eventually recalled having seen a year or so before a painting, an allegory of one of the seasons, that included a group of angels "which seemed to actually float in the air." He also remembered hearing a choir perform an early Mass for Easter morning. Those images and sounds, he concluded, matched his dream, "with the exception of my mother's likeness." And he declared with relief, "Thus everything was explained satisfactorily, and in conformity with scientific facts."[32]

Several years later, Tesla experienced a similar intuition regarding one of his sisters. From New York, he sent home a telegram, "I had a vision that Angelina was arising and disappearing. I sensed all is not well." In fact, she was fatally ill, and a return telegram soon confirmed her death. This time when the scientist tried to explain his precognition, he decided that he simply was a sensitive receiver capable of registering cosmic disturbances that conveyed important news. On a separate occasion, Tesla gave a dinner party for several close friends, but as they were about to depart for a late train to Philadelphia, he had a premonition that "ill adventure would befall them." Tesla delayed them, caused them to miss their connection, only to learn the following day that the train experienced a terrible accident that killed many passengers. Tesla offered no scientific explanation for that forewarning.

In Tesla's life, the line between supernatural and inspirational was thin. While roaming in the mountains after his mother's death, he saw an approaching storm and scrambled for shelter. From his perch, he witnessed a sudden lightning flash followed a few moments later by a deluge. "This observation," he said, "set me thinking." If the lightning and storm were closely related, maybe lightning acted "much like that of a sensitive trigger." That observation led to this insight: "If we could produce electric effects of the required quality, this whole planet and the conditions of existence on it could be transformed. If it were in our power to upset it when and wherever desired, this mighty life-sustaining stream could be at will controlled." And from that insight he committed to developing an even more powerful transformer to provide "electric forces of the order of those in nature." With that "sensitive trigger" resonating upon the earth's inherent energy flows, Tesla predicted he could transmit electricity and messages through the planet without wires.[33]

7.
DIVINE ORGAN OF SIGHT
Chicago

Tesla keenly felt the loss of his mother—his inspiration and key supporter. Yet alongside the personal hardship, he made professional headway on his European trip. Flush with funds from having sold licenses to British and French manufacturers, he returned to the United States in late August 1892 and switched his residence to the posh, eleven-story Gerlach Hotel on 27th Street, between Sixth Avenue and Broadway. Built in the Queen Anne style with bowed bays along the sides and red brick walls trimmed with stone, the Gerlach was "an elegant structure . . . [that] offers all that is possible for luxuriousness in furnishings and delight in cuisine."[1] Tesla's dwelling was referred to as a French flat to distinguish it from most New York "apartments" of the period that were squalid and crowded tenements. (The residential building, now with a few retail shops on the ground floor, has been renamed the Radio Wave Building and features a plaque commemorating Nikola Tesla, who placed a receiver on the hotel's roof in order to capture some of the first radio transmissions from his downtown workshop.)

After arising at 6:30 a.m., having gotten three hours of sleep, Tesla enjoyed a light breakfast, performed a few gymnastic exercises, and began his daily thirty-block walk from his fancy hotel to his new and larger laboratory space on the fourth floor of an uninviting huge yellowish brick building[2] just south of Washington Square and in the heart of Greenwich Village that was teeming with cheap restaurants, wine shops, and weather-beaten tenements.

Tesla counted his steps, making sure they were divisible by three. The journey initially took him by the new Madison Square Garden, a massive indoor arena recently designed by famed architect Stanford White and topped by a gilt-copper statue of Diana sculpted by Augustus Saint-Gaudens. (White would become one of Tesla's close friends, Saint-Gaudens would meet the inventor at dinner parties, and the statue would be moved to the World's Columbian Exposition in Chicago, where Tesla would later organize a grand demonstration of his alternating-current system.) Tesla strode past Madison Square Park, a lovely six-acre public space, and then south on Fifth Avenue until the city reverted to a village. The upper sections of Fifth Avenue were lined by elegant gas lamps and spotted with fancy carriages. As Tesla moved south, horse-drawn carts moved an array of goods from markets to shops, and the street became more crowded and louder with hawkers advancing their wares. Each new block featured stalls promoting different products, including emporiums of buttons and freshly blown glass. As Tesla wove his way through the crowds, the streets became lined with old brick residences of the pre–Civil War period. That journey marked another of Tesla's alternating worlds, having his bed on elegant 27th Street and his head in the village among peddlers.

Most of Tesla's hours were spent in his shirtsleeves or dark frock coat, working diligently with his staff and on his instruments. He only used his small office, furnished with a rolltop desk and a rug, on the rare occasion when guests wanted to talk privately. According to a visiting reporter, that office, reflecting Tesla's personal orderliness, "was immaculate, not a speck of dust is to be seen; no papers litter the desk, everything just so."[3]

Tesla joined the AC versus DC battle in 1891 with his presentation at New York's Columbia College. Now he devoted himself to winning the "War of the Currents" by powering the World's Columbian Exposition scheduled for Chicago in 1893. He and his partner, George Westinghouse, were determined to demonstrate the effectiveness of alternating current at the international fair. Tesla would again be facing Thomas Edison, who touted direct current (DC) for maintaining a constant low voltage all the way from the power generator to the ultimate customer. Focused on safety (and his own profits), Edison argued his DC-based system produced power not strong enough to cause dangerous electric shocks. "We're set up for direct current in America," he declared. "People like it, and it's all I'll ever fool with.... Spare me that nonsense. [Alternating current is] dangerous."[4] Direct current's shortcoming, as noted previously, was that it lacked the force for long-distance transmission.

The different approaches demonstrated different views about electricity's future. Although he launched the Pearl Street Station that transmitted power to nearby offices, Edison tended to think electricity would remain a luxury item enjoyed by those, such as J. P. Morgan, who could afford smaller generators in their homes. Tesla's vision was more democratic, with centralized power plants sending electricity throughout communities. For many years, Edison seemed to be winning, since Pearl Street remained the only central station in 1883 and isolated plants had increased to 334.

Several leading scientists—including Lord Kelvin, Werner von Siemans, and Elihu Thomson—initially feared high-voltage power lines could electrocute innocent bystanders, yet it was Edison who led the charge. Unwilling to consider technologies he did not develop, he had become, according to a biographer, "the stubborn, reactionary old man of the electrical industry."[5] Edison certainly had a clear financial motivation to challenge Tesla's disruptive notion that electricity did not need to flow only in one direction. The Wizard and his allies had sunk millions of dollars into a direct-current system that included generators, wires, and motors. That system was, as the novelist Starling Lawrence put it, "a kind of theology written in steel and copper, and all dedicated to a single false assumption." Edison, he continued, "was shackled to his own success [and] could not jump, or chose not to."[6]

Out of the limelight, Tesla sought to prove the commercial effectiveness of his AC motor and transmission system. L. L. Nunn, manager of the Gold King Mine, needed cheap power for his facility in the rugged San Juan Mountains above Telluride, Colorado. He was running out of nearby wood and coal was not available. He approached the Westinghouse Company in 1891 about transmitting electricity three miles over rough mountain terrain from a generator near a large waterfall. Tesla was called upon to install an AC generator in the valley, string $700 worth of copper wire up the steep hills, and place a transformer at the mine that stepped down the current to run a hundred-horsepower Tesla motor that would power the mine's operations. Much to the delight of the Westinghouse engineers, the system withstood the mountain's frequent storms and high winds, providing reliable power. *Electrical Engineer* declared the Tesla motor a success and that "work in this field is fast passing from experimental investigation into practical electrical engineering."[7]

William Stanley, Westinghouse's chief scientist, had tested similar transformers in Great Barrington, Massachusetts, in 1886. In his small lab, he placed a twenty-five-horsepower steam engine, an alternator, and a transformer that stepped up the voltage from five hundred to three thousand. He then nailed two No. 6 wires along the elm trees bordering the town's sidewalks. In the cellars of six buildings, he placed other transformers that stepped the voltage back down to a pressure that would power lamps. Although the wealthy town already featured a small Edison direct-current power system, Stanley demonstrated a reliable and effective transformer for alternating current. (In 1890, Stanley formed the Stanley Electric Manufacturing Company, in which General Electric in 1903 purchased a controlling interest, and he would thereafter assert that Tesla had stolen his AC ideas from himself and Galileo Ferraris.)

Edison, as noted earlier, tried for years to discredit AC as dangerous, escalating from pamphlets to gruesome electrocutions. Tesla responded by again taking the stage with more shows highlighting the wonders of alternating current. In the battle to obtain the World's Columbian Exposition contract, which promised to settle the "War of the Currents," Tesla gave thought-provoking lectures and demonstrations at the Franklin Institute in Philadelphia in February 1893 and then

in St. Louis in March, but he did not limit himself to the glories of transmitting alternating current long distances. His mind was already rac-
ing to the next challenge.

In Pennsylvania, he suggested, for the first time publicly, that electricity could be transmitted wirelessly. "It is practicable," he said, "to
disturb by means of powerful machines [such as the Tesla coil] the electrostatic conditions of the earth and thus transmit intelligible signals
and perhaps power." Tesla appreciated that this theory would puzzle the
assembled scientists and admitted later that "only a small part of what
I had intended to say [about wireless transmissions] was embodied"
in the speech. Yet it was this "little salvage from the wreck," especially
his principle of utilizing resonance to obtain maximum sensitivity and
selective reception, that later would allow him to claim the title "Father
of the Wireless."[8] In fact, these short references to wireless, only a tiny
portion of the hundred-page lecture, became a key point in later legal
battles over who invented the radio.

In St. Louis, the scientist packed the Grand Music Entertainment
Hall, which seated four thousand but attracted a few thousand more
willing to stand, creating a room "crowded to suffocation."[9] Tesla had
become a celebrity, and a bulletin summarizing his life sold more than
four thousand copies on the streets of St. Louis. Eager attendees, however, admitted the lecture got off to a slow start, with Tesla rambling for
almost twenty minutes about the "divine organ of sight." According to
the scientist, "Of the parts which constitute the material or tangible of
our being, of all its organs and senses, the eye is the most wonderful."
He continued: "The eye is the means through which the human race
has acquired the entire knowledge it possess, that it controls all our
motions, more still, all our actions."[10]

Tesla only slowly drew a connection between the eye and light, and
then proclaimed: "The phenomena of light and heat ... may be called
electric phenomena," which he described as "the mother science of all
and its study has become all important." Foreseeing the future accurately, he predicted: "Power transmission, which at present is merely a
stimulus to enterprise, will someday be a dire necessity."[11]

Audience members became mesmerized by his words as well as his appearance. His jet-black hair covered a rather large head with a high broad forehead. One reporter suggested his wedge-shaped face featured a chin as pointed as an ice pick and a mouth that was too small. Several people noted his big hands and particularly large thumbs; indeed, another journalist suggested these outsized appendages were "a good sign [since] the thumb is the intellectual part of the hand."[12]

The experimenter finally began to entertain the gathering when he brought a metallic ball to within eight or ten inches of his coil's free terminal, provoking "a torrent of furious sparks." When he moved his free hand close to the coil's other terminal, the air became "more violently agitated and you see streams of light now break forth from my fingertips and from the whole hand."[13] As Tesla increased the frequency, the streamers turned from a purplish hue to bright white, and the pungent whiff of chlorine bleach permeated the large auditorium as electric streamers disturbed the air and created ozone. Violet glowworms danced across the walls and ceiling, producing a thick and wavy cobweb of light. To counter Edison's claims about the dangers of alternating current, Tesla offered himself as the subject of the experiment and he assured the nervous crowd "the streamers offer no particular inconvenience, except that in the ends of the finger tips a burning sensation is felt."[14]

As he had done in New York, London, and Paris, Tesla also waved differently shaped tubes that in the presence of a strong electromagnetic field provided "wonderfully beautiful effects . . . the light of the whirled tube being made to look like the white spokes of a wheel of glowing moonbeams." To the delight of the audience members, who shouted "Bravo" at his performance, Tesla held a gas-filled bulb in one hand and touched his other to the oscillating transformer. When the lamp glowed brightly, he recalled, "there was a stampede in the two upper galleries and they all rushed out. They thought it was some part of the devil's work, and ran out. That was the way my experiments were received."[15]

Tesla admitted his glowing tubes and spark-tinged fingertips offered no immediate practical application, yet practicality was not the point. While some displays featured "only a few play things" that were of little value to the great world of science, he was introducing his audience to the future. He was challenging, almost taunting, them to follow his mind.

What was noteworthy in Tesla's St. Louis lecture was his willingness to expand on his dreams for wireless transmissions. Suggesting this ideal "constantly fills my thoughts," he predicted "the transmission of intelligible signals or perhaps even power to any distance without the use of wires." Engaging the audience in his future plans, the inventor said he was working to "know what the [electrical] capacity of the earth is and what charge does it contain if electrified."[16]

Several hundred fans swarmed the dramatic inventor in the Exhibition Theater's lobby, trying to shake his hand. The germophobic Tesla turned pale and tried to move through the crowd quickly. He "had expected a little gathering of expert electricians," reported one newspaper. "Though he went through the ordeal bravely, no power on earth would induce him to try anything like it again."[17]

While Edison's grotesque theatrics might have been more sensational, Tesla's entertaining lectures were more persuasive. Tesla both debunked Edison's claims and demonstrated the widespread benefits available from long-distance alternating current. As a result, Westinghouse began to win the commercial contest, selling AC systems in 1893 that powered twelve times more lights than Edison. Even Edison's own Detroit Edison station manager successfully advanced a resolution at the 1889 meeting of the Edison Illuminating Company to have the parent firm provide "a flexible method of enlarging the territory which can be profitably covered from their stations for domestic lighting by higher pressures and consequently less outlay of copper."[18]

Meanwhile, robber barons were wrestling to control the nation's electrical power. Henry Villard, Edison's largest backer (and the man who drove the golden spike that linked the nation's rail lines), had been working secretly with J. P. Morgan to consolidate the electrical equipment manufacturers. The two of them first approached George Westinghouse, who expressed initial interest in merging with the

Edison General Electric Company, but Thomas Edison would have none of it. "Westinghouse," he said, "has gone crazy over sudden accession of wealth or something unknown to me and is flying a kite that will land him in the mud sooner or later."[19] After Edison won a court judgment for his invention of the incandescent lightbulb (reversing a previous decision), the equally competitive Westinghouse soured on a deal, prompting Villard to reach out to Tesla directly. While Tesla thought a merger would further advance his AC system, he couldn't convince Westinghouse, who owned many of his patents. "I have approached Mr. Westinghouse in a number of ways and endeavored to get to an understanding," Tesla wrote to Villard, "[but] the results have not been very promising . . . Realizing this, and also considering carefully the chances and probabilities of success, I have concluded that I cannot associate myself with the undertaking you contemplate."[20]

Villard turned his merger-making attentions to Charles Coffin, a former shoe manufacturer who had bought the patent rights of Elihu Thomson and Edwin Houston, who had been high school classmates, and became chief executive officer of the Thomson-Houston Company. By offering easy credit and by accepting the securities of local companies generating and delivering electricity, the brilliant and fast-talking Coffin had expanded Thomson-Houston's business and increased its value beyond that of Edison General Electric. A merger of those two firms made both financial and technological sense, in part because the companies possessed complementary patent holdings. Edison General Electric dominated urban DC stations, DC power transmission, and street railways; Thomson-Houston's strength lay in arc lighting and alternating currents. Naming the merged company, however, sparked the most controversy. Thomson and Edison vehemently opposed each other's name in the title. The only solution was to simply call the new firm "General Electric." Although Edison, then forty-five years old, maintained a position on the board, he was stunned and hurt by the change.

At this stage, the "War of the Currents" seemed to have Tesla winning the public relations struggle but General Electric (GE), now without Edison, assembling some serious money. The deciding battle would be waged in Chicago at the World Columbian Exposition of 1893. The international world's fair—marking the 400th anniversary of Columbus landing in the Americas (admittedly one year late in order to accommodate a presidential election)—ultimately attracted twenty-seven million visitors (half Americans, half foreigners). Even in the midst of an economic recession, American boosters wanted to celebrate and highlight their country's engineering superiority with a massive display of electricity's wonders.

More than seven thousand workers spent more than a year converting a bleak swampy bog seven miles south of the city on Lake Michigan into an opulent fairground. The landscape artist Frederick Law Olmsted arranged a Venetian-style system of canals and lagoons surrounding massive exhibition halls and ornate palaces designed by architect Daniel Burnham. The classical Beaux Arts structures suggested a model metropolis, even though Burnham was designing skyscrapers in Chicago and other cities that dramatically altered the urban skyline. Common roof pediment lines and creamy marble integrated this surreal White City, which became the electricians' greatest canvas for their modern wonders.

The new General Electric Company misjudged the proud Chicago organizers who wanted to showcase their raw and vibrant metropolis and did not trust the eastern elite. Sensitive to being confused for rubes, they balked at the sky-high rates initially demanded by the "electrical trust." Although the fair committee had been paying only $11 per lamp to light the construction site for night work, General Electric bid $38.50 per lamp for six thousand lights to illuminate the completed fairgrounds. Chicago organizers worked out a short-term and cheaper deal with several smaller firms, prompting the local papers to herald: "Electric Light Combine Humbled" and "Cannot Rob the Fair."

Seeing an opportunity, Westinghouse took a high-profile gamble and bid just half of General Electric's final quote of $1.7 million to supply and operate the fair's lamps and motors. "There is not much money in the work at the figures I have made," he admitted, "but the advertisement will be a valuable one and I want it."[21] Although Westinghouse won the critical contract, he immediately faced legal and technical challenges. General Electric threatened patent-infringement lawsuits if Westinghouse used, or developed something close to, the Edison incandescent lightbulb, which featured a carbon filament within an all-glass globe from which air was evacuated. Successful litigation, boasted a hard-charging GE executive, would place Westinghouse "entirely in our power. He will not be able to make his own lamps, and he can only buy them from us."[22]

Rather than rely on his own lawyers, Westinghouse turned to Tesla and the company's engineers to develop a unique two-piece lamp, in which a rubber "stopper" could be removed in order to replace burned-out filaments. Other inventors had tried the approach and failed, largely because the design couldn't seem to hold a vacuum, without which the filament would be exposed to air and burn out quickly. Tesla and the Westinghouse team tested scores of combinations, but even the strongest filament within the most reliable "stopper" bulb expired relatively quickly. To maintain a hundred thousand lighted bulbs, therefore, Westinghouse arranged for trains each evening during the fair to transport damaged lights from Chicago to his Pittsburgh factories and others to return in the morning with repaired replacements. The result, despite the effort and expense, transformed the World Columbian Exposition into a glowing wonderland.

Lamps were not Westinghouse's only challenge. What Chicago required was on a scale far grander than anyone had attempted. The Westinghouse-Tesla team upgraded motors and transformers, and they enlarged a dozen generators. What Westinghouse and Tesla assembled became the world's largest AC system, powering the greatest concentration of artificial light, yet, amazingly enough, their thousand-square-foot switchboard required only one engineer to operate the forty circuits that delivered electricity throughout the park.

The effect staggered wide-eyed visitors as well as skeptical engineers. When President Grover Cleveland pushed an ivory-and-gold button on May 1, 1893, the seven-hundred-acre fairground burst into light, elaborate fountains shot water a hundred feet into the air, accompanied by cannon fire, and a choir and orchestra performed Handel's "Hallelujah Chorus."

A shroud fell from Daniel Chester French's giant Statue of the Republic. Children's writer L. Frank Baum claimed the Chicago marvel inspired his Emerald City and the Wizard of Oz. Many visitors wept tears of joy.

The fair, in fact, marked a turning point for the country. At the same time Buffalo Bill's Wild West shows were drawing sell-out crowds, Frederick Jackson Turner, a thirty-two-year-old University of Wisconsin history professor, was lecturing to his American Historical Association colleagues that the frontier had closed. Mass marketing, moreover, was opening opportunities for national brand products, including Cream of Wheat cereal, Aunt Jemima pancake mix, Cracker Jacks, and Juicy Fruit gum. Also previewed at the fair were the zipper and yellow pencils.

The World Columbian Exposition and Tesla's efforts need to be put in some historical context. In addition to displaying the new abundance of mass-produced products, what Louis Mumford would later refer to as "the goods life,"[23] the Chicago fair highlighted a golden age of invention. Electricity, telephones, motion pictures, elevators, automobiles, typewriters (and the list goes on) altered the landscape and daily life, as well as conveyed a sense of progress and possibilities. Historian Thomas Schlereth referred to this period in the late nineteenth century as the "age of modernization" that included "industrial production, a commercial agriculture, technological innovation, the accumulation of capital, an urban consciousness, bureaucratic organization, occupational specialization, and widespread education."[24] Progress, of course, was not linear. Economic depressions—including the Panic of 1893 in which railroad firms collapsed and hundreds of banks failed—slowed its development, but fair attendees felt the world and their lives were changing.

Each evening as twilight loomed within the White City a fabulous spectacle evolved. The gold-domed Administration Building was the first to glow, provoking sighs from the appreciative crowd. Then the columns surrounding the Great Basin burst forth from the growing shadows as thousands of Westinghouse "stopper" lamps illuminated the cornices, pediments, and statues. The crowd clapped wildly. New bulbs then set the palaces alight, offering radiant images of a glorified age.

Arc lights illuminated the walkways, bright bulbs shimmered across the dark waters of the 1,500-foot Great Basin, ghostly gondolas suddenly appeared from their docks, and two massive fountains shot into the air twenty-two thousand gallons of water a minute. As if in a finale, which could be seen eighty-five miles away in Milwaukee, four massive searchlights raked the night sky, switching from dazzling white to light blue, to green, and then to blood red.

The fair's electrical system powered two hundred fifty thousand lamps, scores of cascades, hundreds of exhibits, the Intramural Elevated Railway, the launches that plied the lagoons, the Sliding Railway on the thousand-foot pier (essentially an early escalator), and the great Ferris wheel (rising 262 feet, with the capacity of carrying more than two thousand people, while charging fifty cents for two revolutions). The exposition, in fact, generated and consumed three times more power each day than did the entire city of Chicago, and it was declared "a magnificent triumph of the age of electricity."[25] Just three years before, the Paris Exposition's electrical display had helped earn Paris the nickname City of Light. In comparison, Chicago's Columbian Exposition featured ten times the lamps and horsepower of the French fair.

The Electricity Pavilion itself covered the equivalent of more than two football fields and was as wide as another. The second floor featured electrical devices for curing ailments and promoting personal hygiene, including electrical hairbrushes, charged belts to spur sexual activity, and body invigorators. A model home suggested the domestic future belonged to electric stoves, dishwashers, washing machines, vacuum cleaners, as well as electric doorbells and fire alarms.

The building's main floor displayed the inventions of major companies and inventors, several of whom gave lectures and demonstrations. Elihu Thomson, for instance, presented his own high-frequency coil that sent sparks five feet into the air. Thomas Edison debuted his phonograph (or talking machine) and kinetescope (or moving picture projector); ironically, one of the only Edison incandescent bulbs in the entire fairground was an eight-foot-tall, half-ton model at General Electric's display. The Allegemeine Elektrizitats Gesellschaft (AEG) Company from Germany displayed the alternating-current equipment used in the "epoch-making" 108-mile Lauffen-to-Frankfurt transmission. Yet leaving no doubt about what company won the fair's electrical contract and which inventor controlled the relevant patents, the center aisle of Electricity Hall featured a monument standing forty-five feet tall and spelling out in big bold letters: "Westinghouse Electric & Manufacturing Co. Tesla Polyphase System."

Introduced for his formal lecture on a sultry Friday night by white-haired Elisha Gray, a telephone pioneer, this "Wizard of Physics" appeared to be "a tall, gaunt young man" dressed dapperly in a "neat four-button cutaway suit of brownish gray." A few attendees noticed that his shoes featured thick soles of insulating cork. Tesla smiled but "modestly kept his eyes on the floor." One journalist said "his cheeks were hollow, his black eyes sunken . . . but sparking with animation."[26] Another, however, described his appearance as "somewhat like a resurrected cadaver."[27]

More than a thousand engineers "crowded about the doors and clamored for admittance," most of them hoping to witness high-voltage AC electricity pass through the inventor's body. According to a newspaper reporter, "Mr. Tesla has been seen receiving through his hands currents . . . [that vibrated] a million times per second, and manifesting themselves in dazzling streams of light. . . . After such a striking test, which, by the way, no one has displayed a hurried inclination to repeat, Mr. Tesla's body and clothing have continued for some time to emit fine glimmers or halos of splintered light. In fact, an actual flame is produced by the agitation of electrostatically charged molecules, and the curious spectacle can be seen of puissant, white, ethereal flames that do not consume anything, bursting from the ends of an induction coil as though it were the bush on holy ground."[28]

Tesla's machines offered dazzling flashes that, according to one writer, produced "the effect of a modified lightning discharge . . . accompanied by a similar deafening noise."[29] Across a long table in his own display room the inventor assembled an assortment of coils and motors, providing what another journalist described as "a fair idea of the gradual evolution of the fundamental idea of the rotating magnetic field."[30] Two distinguished professors, when asked by a reporter about the various equipment, "gazed in wonder and confessed they could not guess" the purpose of what they referred to as "Tesla's animals." Phosphorescent tubes and lamps glowed in a darkened alcove. Twisted tubing, the prototype of neon display lighting, radiated "Welcome, Electricians," while other cursive lights honored great scientists such as Faraday and Franklin, as well as Tesla's favorite Serbian poet, Zmaj Jovan. To demonstrate how poly-phase currents create rotating fields, Tesla set ostrich-sized copper eggs twirling rapidly atop a velvet-covered table, under which hummed a four-coil magnet. Electric clocks displayed perfect synchronization.[31]

Tesla, thus, effectively challenged Edison's claims of AC's dangers, demonstrating that high-voltage currents, when frequencies also were high, could be completely harmless. According to one report from the event, "Professor Tesla [sent] a current of 100,000 volts through his own body without injury to life, an experiment which seems all the more wonderful when we recall the fact that the currents made use of for executing murderers at Sing Sing, NY, have never exceeded 2,000 volts."[32]

More importantly, both tourists and scientists came away from the fair convinced the future of energy belonged to electricity and its future belonged to alternating current. They had glimpsed the outlines of a radiant electrical world that would alter how they spent their evenings, how they commuted to work, how machines would take over the arduous tasks long performed by animals and backbreaking human labor. One hundred twenty-five years later, today we realize the Chicago World's Fair illuminated our modern era.

Tesla and Westinghouse had won the "War of the Currents"! After the World's Columbian Exposition, some 80 percent of electrical devices ordered in the United States utilized Tesla's designs.

With this success behind them, Westinghouse and Tesla turned to the enormous hydroelectric opportunities at Niagara Falls, where thundering green torrents fell violently down a 160-foot cliff and made rainbows dance and the earth tremble. Engineers had long assumed the only options to move such power were pulleys, hydraulics, or water-wheels that provided limited service to buildings located no more than several hundred feet from the Niagara River's rapids. To transmit larger amounts of power beyond the small community of Niagara, hopefully to the industrial facilities twenty-two miles away in Buffalo, some scientists suggested sending pressurized water through pipes, while others proposed using compressed air.

The World's Columbian Exposition expanded those options by demonstrating an alternating-current system to be both possible and preferable. Still, Westinghouse and Tesla faced seventeen rival firms that submitted bids to tap the torrent. The aggressive competition even prompted charges of industrial espionage. When Westinghouse suspected that General Electric engineers had stolen his blueprints, he obtained a court order directing the local sheriff to enter GE's Lynn plant. Lo and behold, the sheriff found the missing documents in their possession. GE officials claimed they were just making sure Westinghouse did not pirate their own patented lightbulbs, and a jury split six to six on a verdict. Ever the gentleman, Tesla described the spying controversy as "a trifle, to be sure, not worthy of any consideration."[33]

Deciding among the various bids was Edward Dean Adams, a respected engineer and investment banker with close ties to J. P. Morgan and an indirect lineage to two U.S. presidents. A slight man with a large handlebar mustache, Adams in 1890 assembled 103 financiers, considered "one of the most powerful combinations of New York capitalists ... ever ... formed,"[34] to create the Cataract Construction Company, which then launched the International Niagara Commission to obtain expert engineering advice on harnessing the enormous hydroelectric potential from Niagara Falls.

Westinghouse initially ignored Adams's pleas for a bid, growling: "These people are trying to secure $100,000 worth of information by offering prizes, the largest of which is $3,000. When they are ready to do business, we will show them how to do it."[35] Westinghouse also was buying time for his Pittsburgh-based engineers to devise a more efficient way to wind the AC motor's rotor as well as to design an effective converter that enhanced the long-distance transmission of power—refinements that would allow Westinghouse to make full use of Tesla's poly-phase system. This delay conveniently allowed Adams to witness AC's benefits and safety in Chicago, and he increasingly turned for advice to Tesla, who encouraged the engineer to embrace alternating-current technologies that would electrify and bring economic development to a wider area of New York State. Tesla baldly declared, "It would be entirely impossible for any system to compete (with my) ideally simple one."[36]

Adams kept posing both legal and engineering questions. When asked if a Thomson-Houston patent superseded his, Tesla protested it had "absolutely nothing to do with my discovery of the rotating magnetic field and the radically novel features of my system of transmission of power disclosed in my foundation patents of 1888. All the elements shown in the Thomson patent were well known and had been used long before."[37] When asked about support for a direct-current system from the distinguished Sir William Thomson, whom Queen Victoria had recently elevated to become Lord Kelvin (and was not related to Elihu Thomson), a horrified Tesla declared "how disadvantageous, if not fatal, to your enterprise such a plan would be, but I do not think it possible that your engineers could consider seriously such a proposition of this kind."[38]

Adams eventually agreed that rather than have the rushing waters simply turn a series of small waterwheels, two massive Niagara electrical generators should supply more power than all the central plants then operating in the United States. It was not an easy decision; as he noted, it was based on "faith and hope that electrical engineers could produce apparatus much larger in size than ever had been built and that new types which were then hardly beyond the stage of experiment would prove successful."[39] Adams specifically sought bids to divert the river's turbulent waters into three eight-foot-wide penstocks or pipes and then spin the wheels of gigantic twenty-nine-ton turbines, the world's biggest. Those turbines, in turn, would whirl attached steel shafts that would rotate the two electrical generators up on the project's main floor, which was referred to as "the cathedral of power."

Westinghouse won the key contract in October 1893 with a bold plan to utilize Tesla-designed equipment to capture the power of a diverted Niagara River. General Electric, in large part because of J. P. Morgan's clout, was not left empty-handed, receiving smaller awards to construct the transformers and the transmission line to Buffalo.

To close out the "War of the Currents," Westinghouse threw the switch on November 16, 1896, sending the power of Niagara Falls to run Buffalo's streetcars and illuminate its street lamps. Within weeks, the Tesla-Westinghouse system was electrifying factories and high-temperature electric furnaces that enabled the commercial production of lightweight aluminum and hard carborundum. (About a decade later, *Scientific American* ranked these electric furnaces as the era's greatest invention, noting they allowed the creation of "many absolutely or commercially new products.")[40]

The availability of abundant electricity put the Buffalo/Niagara region on the map, attracting the world's largest concentration of electrochemical firms—producing "acetylene, alkalis, sodium, bleaches, caustic soda, [and] chlorine."[41] Within a few years, ten generators at Niagara Falls were transmitting alternating current some four hundred miles to New York City, where it powered new subways and lit Broadway, which became known as the Great White Way.

Although the Niagara Falls project was not the world's first demonstration of long-distance electricity transmission, it was certainly the most dramatic. Back in 1879, Charles Brush designed water-driven dynamos that sent power several miles to high-altitude gold mines in the Sierra Nevada Mountains, where arc lights enabled miners to work through the night and double their productivity. In 1889, Sebastian Ziani de Ferranti transmitted what was then a remarkable eleven thousand volts over seven miles. Two years later, British engineer C. E. L. Brown of the Oerlikon Company and Russian engineer Mikhail Dolivo-Dobrovolsky from the AEG Company powered a small hundred-horsepower motor at the 1891 International Electrotechnical Exhibition in Frankfurt with electricity generated by a waterfall at a cement factory on the Neckar River in Lauffen 108 miles away. (Brown subsequently acknowledged that they succeeded only by using Tesla's patents.)[42]

These early efforts were limited in scope, yet harnessing the power of Niagara Falls demanded a complex feat on a grander scale. According to Westinghouse, "The conditions of the problems presented, especially as regards the amount of power to be dealt with, have been so far beyond all precedent that it has been necessary to devise a considerable amount of new apparatus. . . . Nearly every device used differs from what has hitherto been our standard practice."[43]

The *New York Times* referred to the project as "the unrivaled engineering triumph of the nineteenth century," and the paper declared, "To Tesla belongs the undisputed honor of being the man whose work made this Niagara project possible." The newspaper's four-column spread, complete with a large portrait of the inventor, further commented: "Perhaps the most romantic part of the story of this great enterprise would be the history of the career of the man above all men who made it possible, . . . a man of humble birth, who has risen almost before he reached the fullness of manhood to a place in the first rank of the world's great scientists and discoverers—Nikola Tesla."[44]

The Westinghouse-Tesla poly-phase system at Niagara Falls marked the dynamic shift from a mechanical era to the age of electricity. Tesla was just forty-one years old when he achieved this lifelong goal, not even halfway through his long life. When prominent businessmen gathered at Buffalo's Ellicott Club to celebrate the Niagara project's inauguration, Tesla's introduction prompted "a monstrous ovation." According to newspaper accounts, "The guests sprang to their feet and wildly waved napkins and cheered for the famous scientist. It was three or four minutes before quiet prevailed."[45]

The acclaimed inventor, who had dreamed as a boy in Lika of tapping the power of Niagara Falls, oddly chose to talk about his sense of inferiority. "Even now as I speak," he claimed, "I shall experience certain well known sensations of abandonment, chill and silence. I can see already your disappointed countenances and can read in them the painful regret of the mistake of your choice." In January 1897, on the night of his great triumph, when he was celebrated for creating history-changing technologies, Tesla's introduction suggested an underlying sense of self-doubt. "These remarks, gentlemen, are not made with the selfish desire of winning your kindness and indulgence of my shortcomings," he continued, "but with the honest intention of offering you an apology for your disappointment."[46]

Here in Buffalo we see Tesla again between two worlds. He's at the top of his game, being applauded for a remarkable and practical electrical achievement. Yet in his head, where he's moved on and is struggling to envision the wireless transmission of power, he's uncertain and feeling inadequate. That conflict between pragmatism and visions will play out regularly as the inventor turned to robots, radio, and the wonders of high-frequency oscillations.

The rambling speech at the Ellicott Club also provided insights into Tesla's work ethic . . . and his disdain for the lazy. While declaring the capture of Niagara's power would "offer a guaranty for safe and comfortable existence to all," he suggested the benefit should not be available to "the greatest criminals of all—the voluntarily idle." Toward the end of his address, he described how invention was "the result of long-continued thought and work." With a flourish, he professed: "With ideas it is like with dizzy heights you climb: At first they cause you discomfort and you are anxious to get down, distrustful of your own powers; but soon the remoteness of the turmoil of life and the inspiring influence of the altitude calm your blood; your step gets firm and sure and you begin to look—for dizzier heights."[47]

Tesla paid lengthy tribute to the engineer as an artist, as well as to the Niagara project's noble purpose. "It was a happy day for the mass of humanity," he said, "when the artist felt the desire of becoming a physician, an electrician, an engineer, or mechanician." This dual approach to electrical science, he argued, "has disclosed to us the more intimate relation existing between widely different forces and phenomena and has thus led us to a more complete comprehension of Nature and its many manifestations to our senses." Attendees had gathered to celebrate the commercial capture of Niagara's power, to revel in the project's profit potential; yet Tesla considered harnessing Niagara a moral advancement that "signifies the subjugation of natural forces to the service of man, the discontinuance of barbarous methods, the relieving of millions from want and suffering."[48]

The inventor, finally, used the occasion to predict more and greater innovations. One of his fondest dreams, for which he claimed "fresh hope," was "the transmission of power from station to station without the employment of any connecting wire."[49] No doubt the assembled owners of the new wires carrying Niagara power to New York City did not appreciate that particular prophecy.

The banquet revealed the joys and wonders of the new electrical age. With fanfare, the *Buffalo Morning Express* observed, "Great are the powers of electricity.... It makes millionaires. It paints devils' tails in the air and floats placidly in the waters of the earth. It hides in the air. It creeps into every living thing.... Last night it nestled in the sherry. It lurked in the pale Rhine wine. It hid in the claret and sparkled in the champagne. It trembled in the sorbet electrique.... Small wonder that the taste was thrilled and the man who sipped was electrified.... Energy begets energy."[50] As giddy attendees emerged from Buffalo's Ellicott Club, within what was then the world's largest office building, "there was not a man who went home did not feel ... the glorious knowledge that the introduction of power had been commemorated."[51]

Even science fiction writer H. G. Wells expressed wonder at the Niagara project. The "softly humming turbines," he wrote, "are altogether noble masses of machinery, huge black slumbering monsters.... There is no clangor, no racket.... All these great things are as silent, as wonderfully made, as the heart of a living body, and stouter and stronger than that.... I fell into a daydream of the coming power of men, and how that power may be used by them."[52]

After being unceremoniously removed from the General Electric Company, Thomas Edison eventually acknowledged Tesla's advancement. At the National Electrical Exposition in New York City in May 1895, the Wizard of Menlo Park observed that the conference's power was coming from far-away Niagara Falls and commented: "To my mind it solves one of the most important questions associated with electrical development." Alexander Graham Bell of telephone fame, also at the New York City gathering, added: "This long distance transmission of electric power was the most important discovery of electric science that had been made for many years."[53]

Tesla clearly had become a scientific superstar, with the *New York Times* declaring the Niagara project "could be no better evidence of the practical qualities of his inventive genius."[54] Another newspaper praised Tesla's spirit as "naturally hopeful.... He looks to a time when power shall be so cheap, so universal, that all labor shall be done by tireless machines and every man's life be thus so much more worth living."[55] Yet amidst the cheers, some still expressed caution about Tesla's ideas. One reporter observed, "His inventions already show how brilliantly capable he is, [but his] propositions seem like a madman's dream of empire."[56]

Tesla himself wanted to move on, largely to return to the comforts and stimulation of his isolated lab and new experiments. He needed to tilt his balance back a bit from admiring a real project, now the status quo, and toward envisioning the future. The tour of the Niagara power plant also prompted another of the inventor's personal quirks, what he called "a curious thing about me." "I cannot stay about big machinery a great while," said this inventor of giant machines. "It affects me very much. The jar of the machinery curiously affects my spine and I cannot stand the strain."[57]

8.

EARTHQUAKES
AND FRIENDS
New York City

Despite the remarkable advances at Niagara, prospects for electricity flickered at the dawn of the twentieth century. Electric motors powered only one factory in thirty since manufacturers were reluctant to abandon their steam-powered, belt-driven systems for what they viewed as unreliable generators. Incandescent bulbs—although they released no smoke, consumed no oxygen, and emitted no unpleasant odors—illuminated only one lamp in twenty as most home owners favored the less expensive and more pleasing glow of gas lamps. Even ten years after the Westinghouse-Tesla achievement, electricity served only 8 percent of American homes.

As Tesla and Edison moved on to explore other ideas, engineers and marketers fought new battles over electricity. Engineers no longer offered revolutionary products (e.g., incandescent lightbulb) or disruptive approaches (e.g., alternating current). Instead, power advances in the twentieth century tended to be incremental, with designers simply providing bigger generators and longer transmission lines.

Particularly significant changes occurred in the marketing of power. Most electrical entrepreneurs in the early twentieth century believed electricity would remain a luxury item and that natural gas companies would continue to power most stoves, water heaters, and furnaces. They feared more electricity customers would simply require the construction of more generators and distribution lines, which might increase costs and decrease profits. The preferred business plan was to expand slowly in niche markets and to encourage wealthy consumers to deploy more lightbulbs.

Was electricity for all or for the wealthy? Would power become a necessity or remain a luxury? Smart money favored small and dispersed generators for mansions over central power stations for the masses. Even with the advance of alternating current and the ability for larger generators to transmit electricity over longer distances, both General Electric and Westinghouse preferred the immediate profits from selling isolated generators over the uncertainties of marketing electricity from centralized generators.

Yet even those dispersed generators presented challenges. J. P. Morgan wanted his refurbished mansion on Madison Avenue to be the first residence to use Edison's lights and generator. However, his neighbor, the aristocratic Mrs. James Brown, was not pleased and initially complained about how her house shook as a result of the dynamo in Morgan's basement. To avoid his neighbor's wrath, the banker ordered Edison to place thick rubber pads under the machinery and to pile sandbags around his cellar walls. Mrs. Brown then declared the generator's fumes were tarnishing her silver. Morgan responded by purchasing a different kind of coal. Meanwhile, his in-house system was hardly reliable. Despite an on-site engineer, the electric lights flickered and even cut off regularly, forcing Morgan and his guests to search in the dark for candles and lanterns. Worse, a short-circuited electric wire sparked and prompted a fire that destroyed the desk and expensive rug in his library.

Despite such problems with small generators, prospects of selling electricity from large power plants seemed even dimmer. General Electric focused on marketing small units to hotels, banks, mills, ironworks, and theaters across the country, while George Westinghouse was winning high-profile contracts to install isolated generators at the St. Louis post office and New York State Capitol.

Samuel Insull developed a different vision—bringing power literally to the people by piecing together a giant electric grid. Edison's personal secretary saw great promise in Tesla's alternating current and he envisioned building larger generators and creating giant electricity-selling monopolies. When the General Electric merger of 1892 ousted Edison, Insull rejected the new firm's $36,000-per-year job in favor of a $12,000-a-year position managing the Chicago Edison Company, one of many struggling electricity-generating firms in the Windy City. Sensing the potential for greater profits, he abandoned managing GE's Schenectady Works and its six thousand employees for running a small Chicago firm that then employed only three hundred.

Insull's genius was in seeing the potential to integrate and optimize the demands of diverse electricity customers. To obtain a nighttime load, he took over the isolated street-lighting generators. To obtain a steady demand throughout the day, he provided special rates to large office buildings and industries. He contributed to Chicago politicians and won the right to be the sole supplier of power to streetcar companies. As a result, his power plants could operate regularly and efficiently, thus reducing his costs as well as his rates, which only increased the demand for his electricity. According to Insull, "Every home, every factory, and every transportation line will obtain its energy from one common source, for the simple reason that that will be the cheapest way to produce and distribute it."[1]

The brilliant marketer inverted contemporary thinking that electricity must be a high-priced luxury item. Noting that rates for electricity at the turn of the century were almost 50 percent greater than those for gas, he cut prices from twenty cents per kilowatt-hour to ten cents in 1897, and he continued every year until prices plummeted to just two-and-one-half cents per kilowatt-hour in 1909.[2]

As Insull attracted more customers, he also convinced General Electric to build larger generators that would replace the size-limited, gasoline-powered, piston-driven engines. In October 1903, GE and Chicago Edison opened the Fisk Street Turbine Station that was powered by burning coal and provided a remarkable five megawatts of electricity. The turbine proved to be an engineering wonder since its blades were the first human-made devices to travel faster than the speed of sound. By 1911, Insull added ten twelve-megawatt turbines at the same site.

In many ways, Insull and Tesla had much in common. Both were immigrants. Both arrived in the United States with little more than an introduction to Thomas Edison, and their initial sessions with the Wizard of Menlo Park proved to be remarkably similar. Both dressed and spoke well, and both found Edison to be dowdy and unkempt. According to the status-conscious Insull, Edison's "appearance, as a whole, was not what you would call 'slovenly:' it is best expressed by the word 'careless.'"³

Insull, however, held different views on inventing than either Tesla or Edison. Tesla appreciated invention for its own sake. For Edison, innovation could occur only when competitive entrepreneurs were fighting for an edge. Insull preferred the security of monopolies. No doubt these utilities electrified the nation but, without competition, they have become risk averse and more reliant on politics than on engineering advances. As a result, even though more than a century has passed, both Edison and Tesla would recognize the basic components of today's electric grid.

Tesla, of course, could not be bothered with marketing electricity or building bigger generators. He decided to capitalize on his newfound fame and focus on high-frequency research. Rather than sell power to the masses, the inventor sought to provide it for free to the entire world.

To attract more publicity and investors, Tesla turned again to Thomas Commerford Martin, the editor of *Electrical Engineer*, who agreed to assemble Tesla's writing and patent applications into a text-book entitled *The Inventions, Researches and Writings of Nikola Tesla*. One reviewer called the publication "a veritable bible for all engineers in the field."⁴ The almost five-hundred-page tome featured chapters on alternating-current motors, single-phase motors, rotating magnetic fields, and poly-phase systems. Martin's compendium received glowing comments in *Book News*, *Physics Review*, and *London Electrician*, and was translated into Russian and German.

Particularly noteworthy and prescient was the book's description of the potential for wireless transmission "of intelligent signals or perhaps even power." Admitting this idea "consistently fills my thoughts and which concerns the welfare of all," Tesla described how wires became unnecessary when a sending device and a receiver are tuned to the same frequency, allowing electrical impulses to jump between them. This is similar to the striking of one tuning fork causing another in the same frequency to vibrate.

Tesla also declared it "practicable to disturb by means of power-ful machines the electrostatic condition of the earth."[5] Why not have an enlarged Tesla coil send a resonating current through the planet, thereby making signals and power available to all, without the need for copper wires? Noting the myth of Antaeus, the wrestling god who derived power from the earth, Tesla claimed the idea of wireless elec-tricity transmission was not novel.

Yet even Tesla recognized the boldness, and bizarreness, of this sug-gestion. He admitted that his friends had encouraged him not to speak in public about wireless transmission for fear "such idle and far-fetched speculations would injure me in the opinion of conservative business-men."[6] Yet in an interview with Arthur Brisbane, a respected reporter with Joseph Pulitzer's New York World, the excited inventor couldn't stop himself and proclaimed: "You would think me a dreamer and very far gone if I should tell you what I really hope for. But I can tell you that I look forward with absolute confidence to sending messages through the earth without any wires. I have also great hopes of transmitting electric force in the same way without wires."[7]

Tesla became increasingly obsessed by the wireless ideal . . . and he was increasingly seen as a dreamer. He focused on resonance and vibra-tion. Studying those concepts would help him understand the earth's electrical capacity and the charge it might contain if electrified. Just as a sound wave can be reinforced or prolonged by the introduction of another vibration at the right frequency, Tesla tested various rates of electromagnetic waves. He actually built a small box—a receiver composed of several capacitors and an electromagnetic relay—that he carried around New York City in order to detect vibrations from the oscillating current produced by the transmitter located in his laboratory. On some occasions, he sensed such a current on the roof of the Gerlach Hotel, about 1.3 miles from his workshop, but the readings were not consistent, a result, as he discovered, caused by slight changes in the speed of the steam engine powering the alternator. So to obtain a more constant speed and avoid unwanted vibrations, he experimented with reciprocating generators, which move a piston up and down, as well as various combinations of a steam engine and an electric generator. Tesla labeled the resulting electric generator an "electro-mechanical oscillator," although some referred to it as "Tesla's earthquake machine."

One unexpected application of the oscillator had to do with the process of elimination, as in bowel movements. It seems when people stepped on a platform connected to this new transmitter they experienced "minute oscillations" that, in the scientist's graphic words, "stimulated powerfully the peristaltic movements which propel foodstuffs through the alimentary channels." At first, the sensation was "as strange as agreeable," but after a time, Tesla and his assistants "felt an unspeakable and pressing necessity which had to be promptly satisfied." Looking for a commercial application, the inventor declared such "mechanical therapy," if used regularly in short doses, eliminated constipation, flatulence, and other stomach ailments.[8] Tesla later touted electricity as a cure for dozens of ailments, from hair loss to depression.

Mark Twain, who had met Tesla at the exclusive Players' Club owned by actor Edwin Booth, was the most famous person to try this vibrating therapy. (Tesla and Twain struck up a friendship after the inventor revealed that the writer's early novels helped him through a sick spell when he was fourteen years old, a disclosure that caused this "great man of laughter [to] burst into tears.")[9] Complaining one day of "a variety of distressing and dangerous ailments," Twain initially felt the vibrations provided relief, yet, despite Tesla's warnings, he stayed a little too long on the platform—and needed to rush to the bathroom. When Twain's sessions were moderated, however, Tesla noted that "in less than two months he regained his old vigor and ability of enjoying life to the fullest extent."[10]

Twain quickly became impressed with Tesla's scientific prowess, writing to a colleague: "I have just seen the drawings & description of an electrical machine lately patented by a Mr. Tesla, & sold to
the Westinghouse Company, which will revolutionize the whole electric business of the world. It is the most valuable patent since the telephone."[11] The two enjoyed numerous dinners together, and Twain spent
many an evening at Tesla's lab reviewing the inventor's latest experiments and comparing notes on a joint venture to develop an automatic
typesetting machine. Twain even appealed to Tesla for the commission
to sell in Austria and England some of Tesla's inventions, particularly
the radio-controlled boat, discussed later, that the writer described as
a "destructive terror."[12]

Tesla's oscillator, when accelerated, could shake buildings. According to the scientist, "I had one of my machines going and I wanted to see
if I could get it in tune with the vibration of the building. I put it up notch
after notch. There was a peculiar cracking sound."[13] Tesla didn't realize
it at the time, but the resonance was triggering an artificial earthquake
that shook nearby tenement flats throughout Little Italy and sent its residents running into the street crying "Madredi Dio" and "Gesu Cristo."
Many in the neighborhood had come from Naples, near Vesuvius, and
feared another disaster. As Tesla continued to increase the frequency
in the early evening of August 11, 1896, dishes rattled, chairs slid across
rooms, and cascading plaster piled up on floors. Out in the street, water
gushed from burst mains and telephone poles swayed.

An oblivious Tesla continued increasing the frequency until everything around him trembled dramatically, the way that a particular
frequency of sound can shatter glass. Inside the laboratory, "suddenly
all the heavy machinery in the place was flying around." The inventor quickly grabbed a hammer and smashed his device, just before the
police arrived and demanded an explanation. "I told my assistants to
say nothing," admitted a deceitful Tesla. "We told the police it must have
been an earthquake. That's all they ever knew about it."[14]

The police visit did not deter Tesla's experimenting. He frequently
took his alarm clock–sized oscillator to construction sites. Placing this

equipment on the steel framework of a ten-story structure, Tesla said, "In a few minutes I could feel the beam trembling. Finally, the structure began to creak and wave, and the steel-workers came to the ground panic-stricken, believing there had been an earthquake." Boyishly pleased the authorities were again called, he casually "took off the vibrator, put it in my pocket and went away." He later boasted, "I could drop the Brooklyn Bridge into the East River in less than an hour."[15]

Several years later, a magazine quoted Tesla suggesting "he could set the earth's crust into such a state of vibration that it would rise and fall hundreds of feet and practically destroy civilization." An accompanying illustration showed the earth broken in two.[16] On another occasion, the inventor asserted he could "split [the earth] as a boy would split an apple—and forever end the career of man." Bragging about his newfound power, he also said: "For the first time in man's history, he has the knowledge with which he may interfere with cosmic processes!"[17] (The Discovery Channel's *Mythbusters* television show in August 2006 tried to demonstrate oscillator-provoked vibrations on an old bridge, but the experimenters measured only a bit of shaking rather than "earth-shattering" effects.)

Tesla's high-speed transmitter evoked less damaging and more practical results. It set the stage, as an example, for a new science of "telegeodynamics" that locates ore and petroleum deposits by studying vibrations reflected from strata far beneath the earth's surface. His oscillating transformer also could concentrate power to melt metals.

Tesla was at the top of his game in the mid-1890s. He was enjoying creative spurts, friendships, and financing. He was, in a word, balanced. For a man who grew up in flux, he finally could envision his ideals while being attached to present realities. When he could stay on that cusp, he was at his best.

Tesla was also enjoying life. Having mastered billiards as a student, to the detriment of his studies, Tesla reveled in snookering his wealthy

New York acquaintances. One night at his old haunt, Delmonico's restaurant, he convinced the manager he "had never played" but had only watched his friends for a little while. According to the manager, Tesla "was very indignant when he found that we meant to give him fifteen points. But it didn't matter much, for he beat us all and got all the money." The manager continued, "It wasn't the money we cared about, but the way he studies out pool in his head, and then beat us, after we had practiced for years. [It] surprised us."[18]

Tesla, despite the pranks, increasingly presented a cultured and respected figure. He spoke several languages and, as one sophisticated acquaintance remarked, "is widely read in the best literature of Italy, Germany, and France as well as much of the Slavic countries, to say nothing of Greek and Latin. He is particularly fond of poetry and is always quoting Leopardi or Dante or Goethe or the Hungarians or Russians. I know of few men of such diversity of general culture, or such accuracy of knowledge."[19] According to a journalist, "I believe he could take up any portion of the second part of 'Faust' and continue the quotation textually page by page."[20] Tesla had become a celebrity even outside cultured and scientific circles. According to the *New York World*, a shopkeeper "lowers his voice when he speaks of Mr. Tesla, as Boston cab drivers used to lower their voices in speaking of [the revered boxer] John L. Sullivan."[21]

Fellow scientists often marveled at Tesla's insights, even if they didn't fully understand them. "His work has been of such an advanced order of electrical application," said one, "that comparatively few have a clear idea as to just what he has accomplished."[22]

During this upbeat period, Tesla was introduced to Robert and Katharine Johnson, well-connected socialites who operated an intellectual salon in their stylish townhouse at 327 Lexington Avenue, about a mile from the Gerlach Hotel in the exclusive Murray Hill neighborhood. Although not wealthy, the Johnsons lived that way, with rooms carpeted in thick Oriental rugs and adorned with a grand piano, harpsichord, and crystal candelabras designed to resemble Easter lilies. They were a handsome couple, with Robert's close-cropped black beard and gold-rimmed glasses and Katharine's red hair and lively manners. By the time Tesla met them, the Johnsons had a teenage son, Owen, and a slightly younger daughter, Agnes.

Robert—an accomplished poet and author (most notably in the 1920s of his well-received memoir entitled *Remembered Yesterdays*)—served as editor of *The Century Magazine*, one of the country's most popular monthly journals. The Johnsons and Tesla enjoyed numerous dinners together, often joined by such luminary writers as Rudyard Kipling and Mark Twain, the naturalist John Muir, the sculptor August Saint-Gaudens, the pianist Ignace Paderewski, and the thespian/comedian Joe Jefferson. Johnson's connections and writings proved to be influential politically, as when he published John Muir's reflections on hiking in Yosemite Valley and helped lobby Congress in 1890 to designate the site as a national park. Robert also persuaded Ulysses S. Grant to contribute a series of articles in *The Century*, entitled "Battles and Leaders of the Civil War," and he edited the first installment of the future president's classic memoir.

Tesla's relations with Robert and Katharine revealed a jovial and relaxed side to the usually intense inventor. At Robert's request, Tesla obtained permission from Zmaj Jovanovich, his favorite Serbian poet whom he considered to be the "Longfellow of that country," for his works to be featured in *The Century*. One evening in the Johnsons' living room, Tesla dramatically recited a translation of the poet's ballad "Luka Filipov," about the Montenegrin battle of 1874 in which the warrior Luka captures an enemy Turk and marches him back to his prince.

> *One more hero to be part*
> *Of the Servians' glory!*
> *Lute to lute and heart to heart*
> *Tell the homely story:*
> *Let the Moslem hide for shame,*
> *Trembling like the falcon's game,*
> *Thinking on the falcon's name—*
> *Luka Filipov.*[23]

From that moment on, Tesla would address Robert as "Dear Luka" and Katharine would be known as "Mrs. Filipov."

Robert and Tesla exchanged several notes a week, some dealing with the editing of a Tesla article or a request for the inventor to review pieces from other scientists. Some letters simply arranged logistics for dinners, although the Johnsons, particularly Katharine, extended many more invitations than the busy Tesla accepted. Katharine issued regular appeals: "Come and shed the radiance of your happy countenance upon us all, especially the Johnsons."[24] Tesla often apologized for being deep within his laboratory work, writing, on one occasion: "I fear that if I depart very often from my simple habits I shall come to grief."[25] After Tesla turned down an invitation to vacation with the Johnsons in Maine, Robert wrote jokingly of Tesla's strict regimentation: "You know it isn't safe for you to get more than three miles away from Delmonico's."[26] On another occasion, Robert gently taunted Tesla for not vacationing by saying the scientist was "joined to your idols of copper and steel."[27]

Robert exchanged news with Tesla about his health: "My buncle [or boil] is on my leg and I walk with much difficulty."[28] And they shared publications, poetry, and ideas, with Robert writing, "I am much touched by your remembrance of me in sending the book on Buddhism. I did not know you were enlisted on that side of the campaign but now when I read it I shall think of you even more frequently than usual— which is by no means seldom, let me assure you."[29]

Tesla liked the Johnsons to schedule dinners with potential investors, whom they jokingly referred to in letters among themselves as "the millionaires." The Johnsons also arranged for Tesla to meet eligible women. Katharine, for instance, wrote: "A very charming girl is to be here who wants very much to meet Mr. Tesla. A real one, I assure you."[30] Robert tried a different tack: "Mrs. Johnson expects you tomorrow evening when our lovely Mrs. Anthony is to be with us."[31] Tesla did seem to enjoy a couple of the introduced ladies, as evidenced by his response to one invitation: "I happen to be free this evening. If you have visitors [ordinary mortals] I will not come. If you have . . . Mrs. Anthony—I will come. Please confirm."[32]

One of Tesla's favorite dining companions was Marguerite Merrington. A "dramatic author" and pianist, born in England but schooled at a convent in Buffalo, she wrote plays and operas and was considered to be "tall, graceful, and charming." The inventor often would ask Katharine about the woman's possible attendance at parties, once writing, "Remember to suggest Miss Merrington—if she would come. I know I could be her victim."[33]

Despite his wooing, Tesla kept these relationships, whether with married or single women, casual and focused on lively conversation. Speaking of the inventor's hesitations around acquaintances of the opposite sex, one colleague said, "I fear he will go on in the delusion that woman is generically a Delilah who would shear him of his locks."[34] Commenting on the tall and handsome scientist, a reporter observed, "Impressionable maidens would fall in love with him at first sight, but he has no time to think of impressionable maidens."[35] Other friends noted that his aversion to germs and touching other people did little to foster intimate relationships. A journalist suggested, "Anyone that talks with him for only a few minutes will get the impression that science is his only mistress."[36] Tesla, moreover, claimed some philosophical justification for his sexual reluctance by referencing the writings of Swami Vivekananda—who won notoriety for lecturing at the Columbian Exposition in Chicago on world religions and the role Vendantic Prana (life force) and Akasa (ether) played in modern science. In addition to declaring that Buddhist theories could be "proved mathematically by demonstrating that force and matter are reducible to potential energy," the inventor claimed the Hindu preacher declared chastity offered a clear path to self-discipline and enlightenment.

Tesla could be almost scientific about his views on marriage. He suggested an artist, musician, or writer could enjoy both matrimony and achievement, "but an inventor has so intense a nature with so much in it of wild, passionate quality, that in giving himself to a woman he might love, he would give everything, and so take everything from his chosen field."[37] On another occasion, he observed that "the thinker is confronted with the problem of perpetuating either the species or the mind."[38] Such reasoning, however, could turn to pathos, as when Tesla admitted: "It's a pity, too, for sometimes we feel so lonely."[39]

Dinners at the Johnson home began at 8 p.m. The slender Katharine, bedecked in a Parisian gown and long white gloves, greeted guests when they arrived. She regaled each with a comment, perhaps about the season, and then led them into the drawing room, where she graciously introduced the new arrival to the other guests, offering kind insights on their common interests. With her light reddish-blond hair flowing, she'd then glide away back to the front door. Some described her captivated eyes as "slightly coquettish" and displaying "a daring sense of play beneath a wistful stare."[40] She exhibited charm, that mysterious combination of tact, poise, sympathy, and manners.

Robert ensured that the servants provided each guest with the drink of his or her choice. The host moved purposefully through the growing crowd, a commanding presence in his rich dark beard, hair dramatically parted in the middle, and pince-nez spectacles. He sported a long black coat but avoided neckties in favor of white starched collars. Even the "mountain man" John Muir dressed for the occasions in a three-piece suit, a gold watch chain dangling from his vest pocket.

Despite the elegant attire of the gathered, theirs was a relaxed salon and laughter quickly filled the space. Children Owen and Agnes made brief appearances and then disappeared upstairs.

The dining room was more structured. Katharine arranged seat assignments and placed name cards at each setting. Her own card was at one end of the long table, where she could direct the servants, with Robert opposite her. On either side of her, Katharine positioned her two favorite gentlemen, which on many evenings were Nikola Tesla and Mark Twain. Other guests were arranged after careful consideration of how conversations could flourish.

Striving to appear wealthy on a limited budget, the Johnsons' dining room featured French gray plastered walls with watermelon pink taffeta curtains and a gilt mirror over the mantel. The drop-leaf wooden table and chairs were painted light green. They did splurge on a pressed tablecloth of white damask and a centerpiece with colorful flowers, even during the winter. A brilliantly polished silver salad fork was placed next to each plate, with the meat fork next and then the fish fork. On the right side of the plate, on which a dinner napkin had been folded flat and square, were the meat knife and the fish knife and then the soup spoon. Katharine made sure Tesla had extra napkins placed behind his plate so he could further sanitize the silverware.

Despite the name cards and six-piece place setting, the Johnsons considered their dinners to be informal so they usually avoided the hors d'oeuvre and entrée, leaving the meal order: soup, fish, roast, salad, dessert, and coffee. The key part of the dinner, from their perspective, was the conversation that on nights with talkers like Twain, Muir, and Tesla wandered wondrously.

The Johnsons and Tesla were contemporaries, with Robert being three years older, and Katharine one year older, than Tesla. They provided each other comfort and protection. Learning Tesla was not feeling well, Robert wrote, "It has been a whole week since we saw you and this time you need cheering up. Come as early as possible and get cheered. We are in a jolly mood and the fire is lighted on the hearth and we only need you."[41] The correspondence reveals a warm relationship, Tesla's most significant outside of his family. In response to a holiday invitation, for instance, Tesla wrote (referring to Katharine by his pet name for her): "I am, as you know, very fond of millionaires, but the inducements you offer are so great that I shall put [them] aside . . . to partake in the splendid lunch which Mme. Filipov will [prepare for] Christmas. I want to be at home—327 Lexington Avenue—with my friends, my dear friends—the Johnsons. If you will prepare a dinner for a half dozen and invite nobody, it will suit me. . . . We shall talk of blessed peace and be merry until then."[42]

Katharine worried herself over Tesla's health. During a heat spell, she begged, "I am so troubled about you. . . . I am haunted by the fear that you may succumb to the heat. . . . Find a cool climate. Do not stay in New York."[43] Robert was more jovial, writing, "The rumor is that you have melted in your laboratory."[44]

Although the men exchanged most of the notes, Tesla enjoyed flirty but innocent exchanges with the ebullient Mrs. Filipov, which Robert seemed to be aware of and referenced in his own letters. Katharine McMahon Johnson was a striking beauty of Irish descent and an accomplished conversationalist. She appeared "poised, with shoulders swept back, her head held erect." While her hair featured a few streaks of gray, she "still exuded an air of youth."[45]

Yet Katharine also was considered selfish, egocentric, and histrionic at times. She had a dominant streak that "not only pulled people to her but manipulated them to her needs."[46] She sometimes viewed the famous scientist as a trophy for her salon, something to impress her friends. In fact, she could be blunt about her social motives. "Another charming lady is to be here who does not believe that you are my friend, that I even know you," she wrote to Tesla. "I wish to convince her that you are on my list."[47]

According to her daughter, Katharine revealed "an Irish personality" and could be "gay and lively and fun loving, but also depressive underneath." When "Katharine would go into one of her moods, [she would] just stay in her room and wouldn't come down even for meals." Agnes also suggested Katharine was attracted to the "imaginative and exciting" Tesla, particularly in comparison to Robert, whom, she said her mother felt was "boring, very formal with old world manners . . . a fine old gentleman." Tesla, Agnes reasoned, "might have brought more gaiety into the house."[48]

Katharine might demand Tesla's appearances, with something like: "Dear Mr. Tesla, I shall expect to see you tomorrow evening."[49] Or she could be pleading: "Will you come to see me tomorrow evening and will you try to come a little early. . . . I want very much to see you and will be really disappointed if you do not think my request worthy [of] your consideration."[50] After a few rejections, she could sound desperate, as with: "You do not need anybody, inhuman that you are; how strange it is that we cannot do without you."[51] She also teased, declaring her desire to be "hypnotized" by Tesla's presence.[52] She even complained about competing with her husband for the inventor's attentions, writing: "Why do you not come to see ME instead of always dropping into *The Century* to see

Robert? . . . I must have done something to offend you, but what? How can you be indifferent to such devotion?"[53]

Tesla often demurred. "I cannot come tomorrow," he once explained, "because I already accepted an important engagement with an English Millionaire. I can think of you and rejoice—knowing that you are having an agreeable evening."[54] After apologizing on another occasion for not being able to drop by, he offered his more typical excuse: "My head is full of thoughts on some new inventions and I do not dare to interrupt them for fear of losing the thread."[55] Tesla understood Katharine's displeasure with his excuses and recognized her temper; after demurring about a Christmas dinner invitation, for instance, he wrote to Robert: "Your charming wife is evidently displeased. If she were not Irish I would remonstrate, but under the circumstances I can do no more than wait till her ire is spent."[56]

To compensate for his absences, Tesla would invite the Johnsons periodically out for nights on the town, such as the gala performance of Dvorak's *New World Symphony*. "Nothing better than the 15th row!" he lamented in an invitation to Robert. "Very sorry, we shall have to use telescopes. But I think the better for Mrs. Johnson's vivid imagination. Dinner at Delmonico's."[57]

Tesla greatly enjoyed Katharine's company and their banter. He usually addressed her formally, often as "Mrs. Johnson" or "Mrs. Filipov," but on a few occasions it was "Cher Mademoiselle Johnson,"[58] when writing his letters in French, or "my dear Kate."[59] According to a mutual friend, Katharine's "spell is now a potent one, I fancy, with him, so far as any woman's can be, next [to] his sisters."[60] Yet nothing seemed to be hidden from Robert. Once when Katharine sent flowers to Tesla commemorating the Serbian Christmas, Tesla responded to her husband: "I have to thank Mrs. Johnson for the magnificent flowers. I have never . . . received flowers [before], and they produced upon me a curious effect."[61]

The Johnsons frequently visited Tesla's laboratory, which they described as the "magician's den" where, according to Robert, "lightning-like flashes of electrical fire of the length of fifteen feet were an every-day occurrence, and his tubes of electric light were used to make photographs of many of his friends as a souvenir of their visits."[62] Johnson and Mark Twain were among the first to be photographed with phosphorescent light, and Robert became "the medium of the passage of an electric current of a million volts of the Tesla system of high frequency, whereas I believe twenty-five hundred volts of the ordinary current is sufficient to kill. Lamps were thus lit brilliantly through my body."[63]

Robert even wrote and published a poem entitled "In Tesla's Laboratory":

Here in the dark what ghostly figures press!—
No phantom of the Past, or grim or sad;
No wailing spirit of woe; no specter, clad
In white and wandering cloud, whose dumb distress
Is that its crime it never may confess;
No shape from the strewn sea; nor they that add
The link of Life and Death,—the tearless mad,
That live nor die in dreary nothingness:
But blessed spirits waiting to be born—
Thoughts to unlick the fettering chains of Things;
The Better Time; the Universal Good.
Their smile is like the joyous break of morn;
How fair, how near, how wistfully they brood!
Listen! That murmur is of angels' wings.[64]

As Tesla did with journalists and possible investors, he treated the Johnsons to his constant financial hopefulness. "My new machine will be a wonder," he wrote in a typical letter. "There are millions in it."[65] In another note the inventor declared his talent "can be transformed into car-loads of gold. I am doing that now."[66] Tesla also declared, "You will be pleased to know that I am successfully developing my undertakings and the time for that notice from my bankers [about my new wealth] is near at hand."[67]

In addition to articles by Johnson in *The Century*, Tesla in the mid-1890s was featured regularly in the *New York Times*, *New York Herald*, and *New York World*, as well as regional papers—such as the *Detroit News* and *Savannah Morning News*—across the country. The inventor sought such attention, yet he tried to convey modesty. "It is an embarrassment to me," he claimed to one reporter, "that my work has attracted as much public attention, [in part] because I believe that an earnest man who loves science more than all should let his work speak for him."[68]

Still, awards followed the publicity. The Franklin Institute in Philadelphia in 1893 bestowed upon him the Elliott Cresson Gold Medal, and Columbia University and Yale University followed with honorary doctorates.

Notoriety also attracted some financing. Edward Dean Adams, the engineer and investor who organized the syndicate of Wall Street banks that underwrote the Niagara Falls hydroelectric project, now joined forces with Tesla. In February 1895, they formed the Nikola Tesla Company to "manufacture and sell machinery, generators, motors, electrical apparatus, etc."[69] The firm appeared to have bright commercial prospects with Tesla's technical advances on cold light (fluorescent bulbs), high-frequency oscillators, and radio transmissions.

9.
FIRE AND ROBOTS
New York City

The good times would not last. Adams was far too busy with other projects to provide the critical business guidance that Charles Peck once offered to the quixotic Tesla. Even during their infrequent encounters, the financier and the inventor did not bond. Adams was not a modest man and his biographer claimed the financier had helped "struggling geniuses who afterward saw [with] clearer eyes and performed with greater zeal and skill because they took counsel with Edward Dean Adams."[1] Tesla, as a result, was becoming untethered from realism; he lacked checks on his envisioning.

Unfortunately, Tesla also suffered from bad timing. With the world's economies struggling to emerge from the Panic of 1893 and the United States on the verge of bankruptcy, investors were extremely cautious toward any new venture. More critically, Tesla could not stop dreaming. Rather than manufacture practical products like more efficient lamps, his passion remained inventing, and he was becoming fixated with transmitting messages and power through the earth.

No doubt Tesla's genius was his ability to view challenges from new angles. While Marconi, Hertz, and other wireless forerunners were trying to send electromagnetic waves through the atmosphere, Tesla looked to the ground. Describing Tesla as a maverick thinker, one technology historian observed, "Just as in an AC motor he had decided to have the magnetic field change in the rotor rather than in the stator, now he decided that the ground current should transmit energy and the electromagnetic waves should simply serve as the return needed to complete the circuit."[2]

So while Westinghouse was completing the Niagara Falls project in the mid-1890s, Tesla was asking: Why not take advantage of the earth's natural resonant frequency to propel electricity everywhere on the planet? Why not pump a triggering amount of power into the ground in order to transmit electricity, sound, and data around the world? Why not send waves of current into the earth, have a receiver extract them, and then transmit electromagnetic waves through the atmosphere to complete the circuit? Tesla realized he would need a more powerful oscillator.

What he really needed was rest. The mental strain of inventing and fundraising was showing, with one reporter observing the "pallid, drawn, and haggard appearance of [his] face."[3] Tesla admitted, "I am completely worn out, yet I cannot stop my work. These experiments of mine are so important, so beautiful, so fascinating, that I can hardly tear myself away from them to eat, and when I try to sleep I think about them constantly. I expect I shall go on until I break down altogether."[4]

He was tipped over the edge suddenly on March 13, 1895, beginning at 2:30 in the morning when a fire swept through his Fifth Avenue laboratory. Although not injured, having been asleep at his hotel, Tesla lost everything, including his equipment and notes. The uninsured scientist was, according to *Electrical Review*, "utterly disheartened and broken in spirit. . . . Nikola Tesla, one of the world's greatest electricians, returned to his rooms in the Gerlach yesterday morning [after reviewing the damage] and took to his bed. He has not risen since. He lies there, half sleeping, half waking. He is completely prostrated."[5] Tesla told the *New York Times*, "I am in too much grief to talk. What can I say? The work of half my lifetime, very nearly; all my mechanical instruments and scientific apparatus, that has taken years to perfect, swept away in a fire that lasted only an hour or two. How can I estimate the loss in mere dollars and cents? Everything is gone."[6]

The fire had started in the downstairs storeroom for Gillis & Geoghegan, manufacturers of steamfitter supplies, which often were soaked in oil. One newspaper suggested a careless night watchman, named Mahoney, discarded a still-burning cigarette, and although he tried to douse the blaze using pails of water, the oil-soaked materials burned like a tinderbox. Firemen made little headway against the intense heat, which destroyed the six-story building and even damaged the nearby elevated railroad.

Examining the stark devastation, a reporter described the scene: "Two tottering brick walls and the yawning jaws of a somber cavity aswim with black water and oil were all that could be seen [that fateful] morning... of a laboratory which to all who had visited it was one of the most interesting spots on earth."[7]

Charles A. Dana, then managing editor of the *New York Tribune*, suggested the loss of Tesla's temple of invention had international ramifications. "The destruction of Nikola Tesla's workshop, with its wonderful contents, is something more than a private calamity," he wrote. "It is a misfortune to the whole world. It is not in any degree an exaggeration to say that the men living at this time who are more important to the human race than this young gentleman can be counted on the fingers of one hand; perhaps on the thumb of one hand."[8]

The fire's greatest misfortune, however, might have been to delay Tesla's efforts, particularly on the wireless transmission of intelligence, or radio. In the race to discovery, this postponement afforded Guglielmo Marconi an opportunity to advance unchallenged his devices and patents.

Tesla suffered debilitating depression, which he battled, appropriately enough, through electricity or, more specifically, electrotherapy. It was not unusual at the time to deploy mild shocks to treat an array of ailments—"promote heart action and digestion, induce healthful sleep, rid the skin of destructive exudations, and cure colds and fever by the warmth they create"—but Tesla accelerated the treatments over several months in order to avoid "sinking into a state of melancholia." Each morning he would disrobe, stand naked upon his "vitality booster," and administer higher and higher doses. He later declared electricity "puts into the tired body just what it most needs—life force, nerve force. It's a great doctor, I can tell you, perhaps the greatest of all doctors."[9] Claiming credit for the procedure's development and popularity, he bragged, "Leaders in the profession have assured me that I have done more for humanity by this medical treatment than all my other discoveries and inventions."[10]

Despite the staggering setback and mental strain, Tesla struggled back. Don't forget that Tesla had endured depression and cholera some twenty years before. Showing compassion, Edison offered his own laboratory temporarily to the distraught scientist, and within two months of the fire, Tesla was ordering new equipment from Westinghouse. (Ironically, it was Edison, Tesla's frequent competitor, who offered space for free while the Westinghouse Company, which had profited so much from Tesla's inventions, began to bill him for equipment lost in the fire and to charge him for new machinery.) "I am so very anxious to begin anew my interrupted experiments," he wrote to Pittsburgh officials. "This kind of work is almost essential to my health, and I hope that its resumption will have a good effect upon me."[11] Over the next few months, the inventor kept up a regular stream of equipment replacement orders, usually requesting haste. In one letter, he wrote: "Will you kindly forward, as speedily as possible, rings of sheet iron, 9-1/2 inches inside, and 11 inches outside diameter. The thickness of the sheet iron should be between 1/64 and 1/32 of an inch, or whatever thickness you use in your armatures. The iron should be of the highest quality and best annealed, and I shall want about 100 pounds of these plates."[12]

Tesla expressed flexibility, and Westinghouse offered sympathy. When ordering a motor, for instance, the scientist said, "I may use a machine of say 110 volts if one should be in stock. Furthermore, I am not limited exactly as to the capacity, and it might be something more than 25 H.P., but not very much more."[13] Company officials responded, "The order will receive our best attention." A Westinghouse vice president compassionately added, "I have regretted every time I have thought of your misfortune that you should have met with such a loss. I am very glad to know that you took off your coat and went to work again."[14]

As he rebuilt his workshop, Tesla also rebuilt his relationship with George Westinghouse, who was losing control of his company to bankers who felt far less generous to Tesla. The businessman invited Tesla to come again to Pittsburgh to "observe the operation of your motors in daily practice as well as to get a comprehension of our facilities."[15] The two exchanged regular notes about job seekers or about logistics for meeting representatives of the U.S. Navy or Prince Albert of Belgium. Tesla increasingly was confident and bold enough to suggest strategies to protect his patents and bolster his reputation: "I am prompted by a feeling of friendship to point out to you personally that your advertisements do not bring out clearly the fact that these patents cover not one, but a number of lines of almost equivalent importance. . . . You can readily see that with a few words your statements may be rendered much stronger, and will more exactly express the state of things."[16] The ever diplomatic Westinghouse responded quickly: "We are anxious to give as much prominence as possible to every feature in connection with your inventions."[17]

The equipment arrived at a new workshop Tesla rented on two floors of 46 East Houston Street, located just east of Greenwich Village. While focused on wireless transmissions, the inventor kept his eyes open for new opportunities. Slightly before Wilhelm Roentgen in December 1895 announced his discovery of X-rays, Tesla witnessed the effects of electrical discharge tubes on film, but he didn't recognize the significance of this new form of energy. "Too late," Tesla lamented later. "I realized that my guiding spirit had again prompted me and that I had failed to comprehend his mysterious signs."[18]

After reading Roentgen's articles, Tesla quickly repeated and expanded on the German physicist's experiments. It's possible that Tesla took the first X-ray image, this one of Mark Twain's hand. Such "black light and very special radiation," Tesla said, fashioned "Shadowgraphs" on plates inside metal containers. In one exposure of forty minutes, he revealed "not only the outline, but the cavity of the eye, the lower jaw and connections to the upper one, the vertebral column, and connections to the skull, the flesh and even the hair."[19] Initially dismissing the dangers, he came to realize that radiation produced bad headaches for himself and severe blistering and skin inflammation for an assistant.

Tesla also devised several innovative devices to produce X-rays, and he claimed his instruments would "enable one to generate Roentgen rays of much greater power than obtainable with ordinary apparatus."[20] Several years before observations by quantum physicists, he also suggested insightfully that these powerful energy rays had both particle-like and wave-like properties. X-rays, he further observed, provided a wonderful "gun to fire, projecting missiles of a thousand fold greater penetrative power than that of a cannon ball, and carrying them probably to distances of many miles."[21]

Thomas Edison also came to recognize the usefulness of this new-found energy, and he focused particularly on how X-rays could cause the blind to "see a light." Tesla expressed skepticism of that claim, asking, "Is it not cruel to raise such hopes when there is so little ground for it? What possible good can result?"[22] Reporters responded by trying to provoke another battle between the electrical wizards, with headlines suggesting "Tesla Opposes Edison."[23] The argument, however, was short-lived as the Kentucky School of Medicine combined their devices to examine the birdshot in a wounded person's foot, finding "every bone was distinctly shown, and the shot, about thirty in number, were plainly located." The sometimes combatants and sometimes friends celebrated by enjoying a day of fishing, sponsored by the Safety Insulated Wire and Cable Company, on a topsail schooner off Sandy Hook. According to one press report, they were "as happy and well satisfied a party as ever rode the waves of the Atlantic's billows," with Tesla nabbing "a flounder of large dimensions [and] Edison caught a shocking big fluke."[24]

Recognizing that numerous other scientists and well-financed companies were beginning to dabble with X-rays, Tesla returned to concentrate on wireless transmissions. It was during this period in late 1896 and early 1897 that Westinghouse flipped the switch at the Niagara Falls generating station and Tesla received praise at Buffalo's Ellicott Club. Tesla's mind, however, was elsewhere as his wireless efforts opened a totally new field that he called "telautomatics" (which we now refer to as "robotics"). He predicted electromechanical devices could be regulated from a distance without intervening wires and, more remarkably, to even think for themselves.

The inventor's initial project was a radio-controlled model boat, measuring approximately four feet in length and three feet high, which could receive electromagnetic signals from a transmitter. Other engineers had devised rudimentary wireless systems that transmitted in only one frequency, yet Tesla developed an extraordinary disk that could distinguish among multiple signals and direct an independent boat to flash its lights, rotate the rudder, as well as engage the propeller. "The mechanism found in this invention is among the most sophisticated devices Tesla created in his career," declared a scientific historian. "And up to 1897, no one had conceived of using electromagnetic waves to operate an unmanned vehicle; Tesla introduced the idea into popular culture and engineering practice."[25]

Ever the showman, Tesla arranged for the telautomaton's debut at the massive Moorish-style Madison Square Garden during the Electrical Exposition in May 1898, three months after the Spanish sank the battleship *Maine* in Havana Harbor. The inventor had erected a small indoor pond and invited scientists, possible investors, and the general public, whom he tantalized with progressively entertaining demonstrations. After directing the battery-powered boat to motor around the pool, moving it backward and forward, causing it to dance like a water bug, he asked mesmerized audience members to ask the vessel a question. When one wanted to know the cube root of sixty-four, Tesla quietly flicked levers on a small box hidden next to his podium that made the boat's lights flash four times. "It created a sensation," he remarked, "such as no other invention of mine has ever produced."[26]

Shocked witnesses to Tesla's demonstration assumed the iron-hulled model moved by magic, telepathy, or, according to one observer, a trained monkey hidden inside the small vessel.[27] Even government officials expressed skepticism. When Tesla tried to offer the invention to a military officer, "he burst out in laughter.... Nobody thought then that there was the faintest prospect of perfecting such a device."[28] Moreover, after reading Tesla's startling patent submission, entitled "Method of and apparatus for controlling mechanisms of moving vessels or vehicles," the skeptical chief examiner traveled to New York in November 1898 to see the invention for himself... and quickly approved the application.

Using military terms in order to attract a Navy contract, Tesla admitted his next telautomaton, what he called a "torpedo boat without a crew," sounded "almost incredible." Nonetheless, he declared, "My submarine boat, loaded with its torpedoes, can start out from a protected bay or be dropped over a ship's side, make its devious way below the surface, through dangerous channels of mine beds, into protected harbours and attack a fleet at anchor.... Through all these wonderful evolutions it will be under the absolute and instant control of a distant human hand on a far-off headland, or on a war ship whose hull is below the horizon and invisible to the enemy."[29] The inventor also devised a loop antenna that could be enclosed in the vessel's hull, allowing the boat to submerge, become invisible, yet still receive signals. He filed for patents in thirteen countries.

According to the *New York Sun*, the small robotic boat would "revolutionize warfare."[30] Tesla, ever the optimist, claimed, "Its advent introduces into warfare an element which never existed before—a fighting machine without men as a means of attack and defense. The continuous development in this direction must ultimately make war a mere contest of machines without men and without loss of life."[31] In an article entitled "Torpedo to Revolutionize Warfare," he is quoted as saying, "We shall be able, availing ourselves of this advance, to send a projectile at a much greater distance; it will not be limited in any way by weight or amount of explosive charge; we shall be able to submerge it at command, to arrest it in its flight and call it back, and to send it out again and explode it at will; and more than this, it will never make a miss."[32]

The telautomaton, Tesla alternatively proposed, "will tend to bring about and maintain peace among nations." Through demonstrations and interviews, the inventor described how his tub-like boat could destroy battleships, but he claimed to have "no desire that my fame should rest on the invention of a merely destructive device." He announced, "I prefer to be remembered as the inventor who succeeded in abolishing war." Armed battles, he suggested, "will cease to be possible when all the world knows tomorrow that the most feeble of nations can supply itself immediately with a weapon which will render its coast secure and its ports impregnable to the assaults of the united armadas of the world."[33]

Tesla's success was further highlighted by Marconi's failure. The Italian had devised his own wireless detonation system. In a separate pool he placed a model frigate that was to blow up when it came near a "Spanish ship." Unfortunately, Marconi had not figured out how to send individualized messages on different frequencies so when his assistant—actually Thomas Edison's son, Tom Junior—pushed a button on his control box, an explosion rocked a back room that stored Marconi's miniature bombs, sending shocks and smoke throughout the exhibition hall.

Tesla's imagination, not surprisingly, envisioned something grander than just a robotic boat or weapon. "Had I accomplished nothing more than [a torpedo steered without wires], I should have made a small advance indeed," he said. "But the art I have evolved does not contemplate merely the change of direction of a moving vessel; it affords a means of absolutely controlling, in every respect, all the innumerable translator movements, as well as the operations of all the internal organs, no matter how many, of an individualized automaton."[34] When a reporter called the boat a weapon of war, Tesla sternly countered: "You do not see there a wireless torpedo. You see there the first of a race of robots, mechanical men which will do the laborious work of the human race."[35]

Even Tesla's U.S. patent application foresaw broad applications, noting that the wireless device could control "boat, balloon, or carriage"; be operated by direct conduction through the earth or by electromagnetic radiation through the air; and be used "for other scientific, engineering, or commercial purposes." From our modern perspective, the telautomaton combined advancements in wireless transmission and selective tuning and represented one of the great technological achievements of the late nineteenth century. By using combinations of radio waves in different frequencies, something Tesla called "individualization," he could coordinate tuning devices in order to transmit multiple signals and prevent interference—a major advancement over Marconi's efforts. As a result of Tesla's work, a sender could be confident a message would be obtained only by the intended receiver and would also be decipherable amidst the cacophony of unrelated radio traffic.

Tesla obtained this individualization by placing two circuits within the boat and tuning and combining them "so as to cause the operation of the controlling mechanism when both of these circuits were energized by their respective vibrations" from his oscillating transformer.[36] As a result, he could generate scores of separate frequencies and control multiple vessels simultaneously.

For Tesla, the robot further characterized a mechanical creation that could think. Independent of any operator and left entirely to itself, he said, a future telautomaton would be able "to perform, in response to external influences affecting its sensitive organs, a great variety of acts and operations as if it had intelligence."[37] Going one step further, the robotic boat represented to Tesla the first nonbiological form of life, a machine "embodied" with a "borrowed mind"—his own. Forseeing artificial intelligence, the telautomaton not only would be able "to follow a course laid out or . . . obey commands given far in advance, it will be capable of distinguishing between what it ought and what it ought not to do."[38]

Electrical Review proclaimed the radio-controlled boat to be one of the "most potent factors in the advance of civilization of mankind." Yet such bold pronouncements, even if accompanied by a rather remarkable working model, did not impress some of Tesla's scientific contemporaries. Cyrus Brackett, a Princeton physics professor, scoffed, "There is nothing new about this. The theory is perfect, but the application is absurd. . . . Do you suppose that in the din of battle it would be possible to put into execution those minute and carefully adjusted mechanical experiments, all of which are presupposed by his theory, which require the quiet of an uninterrupted laboratory to work successfully?"[39] Amos Dolber of Tufts College grumbled, "He is getting to be like the man who called 'Wolf!

wolf!' until no one listened to him. Mr. Tesla has failed so often before that there is no call to believe these things until he really does them."[40]

Even his friend Thomas Commerford Martin at the *Electrical Engineer* was growing weary of Tesla's unfulfilled claims for cold lighting, radio, and wireless power transmissions; he complained, "Within the last year or two Mr. Tesla has, it seems to us, gone far beyond the possible in the ideas he has put forth, and he has today behind him a long trail of beautiful but unfinished inventions."[41] Martin's dismissal of Tesla's latest work was no doubt aggravated by the editor's mounting interest in other inventors. Needing new and dramatic stories for his magazine, Martin became particularly intrigued with the wireless efforts of Marconi. In fact, he pushed to have Edison's son, Tom Junior, become Marconi's representative in the United States, an effective way for the Italian to obtain more credibility for his patents and strengthen his claim to being the true wireless pioneer.

The growing strain was not only personal but also financial. Martin's 1893 compilation of Tesla's speeches and patent applications had been a relatively popular publication, but the inventor promptly borrowed and never returned Martin's profits, causing the journalist to complain that "two years of work went for nothing."[42] Tesla, who could get testy, did not appreciate such criticism. Demanding "a complete and humble apology," he huffily wrote: "On more than one occasion you have offended me, but in my qualities both as Christian and philosopher I have always forgiven you and only pitied you for your errors. This time, though, your offense is graver than the previous ones, for you have dared to cast a shadow on my honor."[43]

Although some of the carping can be attributed to competitive and envious scientists, as well as to growing conflicts among rival journals, even Tesla recognized the truth within some of the criticism. He embraced being an idealist and declared, "I want to go down to posterity as the founder of a new method of communication. I do not care for practical results in the immediate present." More telling, Tesla did little to push government or millionaire investors to advance his remote-control electronics and multichannel broadcasting systems.

Undisturbed, the inventor saw his role as revealing principles while leaving it to others to develop the associated applications.[44]

Not long after introducing his robot-controlled boat, Tesla was introduced by the Johnsons to Richard Pearson Hobson, a Naval Academy graduate, maritime engineer, and a southerner from a prominent family who had worked for the Office of Naval Intelligence. Hobson also had experience with naval warfare. In early June 1898, near the beginning of the Spanish-American War, the young lieutenant drove the frigate *Merrimac* on a "dash" into Santiago Harbor under "a lively cannonade of fire" in order to attack the Spanish armada. According to a newspaper account, "Everyone is astounded at the audacity of the American vessel."[45] Hobson, it seems, deliberately scuttled his vessel in order to lock Admiral Cervera and his fleet in the harbor. Although captured and imprisoned in the dungeon of Morro Castle, Hobson became an instant celebrity, with another newspaper proclaiming: "In a day, in an hour, the potent, all-pervading force of electricity . . . flashed his fame over the round world."[46] (Several years later, Hobson received the Medal of Honor for his bravery.)

The famous and muscular officer—striking in his uniform and possessing deep-set eyes, prominent chin, thick beard, and handlebar mustache—often shared dinner with Tesla at the Johnsons', and the two exchanged regular notes, with Hobson in one suggesting: "Now, my dear fellow, if you are doing nothing over the next ¾ of an hour come over for a short tête-à-tête—I feel I have not seen half enough of you on this visit and I have so much to talk with you about."[47]

Tesla even bantered with the Johnsons over access to the good-looking military hero. The scientist wrote to Robert using their pet names: "Remember, Luka, Hobson does not belong to the Johnsons exclusively. I shall avenge myself on Mme. Filipov by introducing him to Mme. Kussner and somebody will be forgotten."[48] He separately teased Katharine: "We must exhibit Hobson. I know he likes me better than you."[49]

Tesla found Hobson to be "a fine fellow" who possessed a keen mind.[50] The two spent "delightful" hours together in the laboratory and at the movies. For Tesla, Hobson offered male companionship, perhaps a substitute for Anthony Szigeti, while Hobson enjoyed the intellectual exchanges about boats, science, and wireless transmissions.

◎ ◎ ◎

In response to the criticism from a few journalists and scientists, the lack of commercial prospects for his automatons, and the encouragement of Hobson and other friends, Tesla decided to return to focusing on his one big idea—wireless transmissions. Taking to heart Lord Rayleigh's recommendation in 1892 to address something grand, Tesla rejected the scientific consensus about Hertzian or electromagnetic waves moving through the atmosphere and doubled down on sending messages and power by electrostatic forces [referring to static electric charges], preferably through the earth, which he claimed to be the "ideal method of transmission."[51]

Tesla could display a competitive spirit when it came to this grand ideal. Comparing his wireless approach to the Hertz-wave system, he said, "The first enables us to transmit economically energy to any distance and is of inestimable value; the latter is capable of a radius of only a few miles and is worthless." (He had made similar distance comparisons between his alternating-current and Edison's direct-current arrangements.) Although claiming not to have the "slightest ill-feeling" toward other experimenters, he snidely suggested "how much better it would have been if the ingenious men . . . had invented something of their own instead of depending on me altogether."[52]

Achieving Tesla's goal of wireless transmission required raising even further the voltage and frequency from his coils. Within the Hudson Street laboratory, he increased his system's output to a remarkable four million volts, which generated stunning sixteen-foot sparks. By early 1895, he also detected oscillations from his lab-based transmitter at West Point, more than fifty miles up the Hudson River, a noteworthy achievement. Yet he still didn't fully understand how to create "a transmitter of adequate power" that might "bridge the greatest terrestrial distances."[53] To address that challenge, Tesla needed to leave the laboratory, "to go into the open" and build a larger project.

10.

LIKE A GOD CONTROLLING NATURE'S POWER

Colorado Springs

n his search for more space to generate bigger lightning bolts, Tesla asked for advice from his patent attorney, whose law partner, Leonard Curtis, suggested Colorado Springs, about sixty miles south of Denver. Curtis, who had moved to the area a few years before in order to enjoy its crisp air and dry climate, argued the mile-high town's thin atmosphere would be particularly conducive to the wireless transmission of power and messages. A bit of a local booster, Curtis enticed the famous scientist with free power at night from El Paso Power Company, the local utility with which he was associated.

Curtis recommended Knob Hill—a slight rise within a cow pasture on the city's eastern outskirts, not far from the Colorado School for the Deaf and Blind—for a sixty-by-seventy-foot barn that would serve as an office and experimental station. He arranged for a local carpenter to design a telescoping mast that could raise a thirty-inch-diameter, copper-covered ball up to 142 feet, a twenty-five-foot tower to stabilize the mast, as well as a retracting roof that would help prevent the laboratory from catching on fire when Tesla's electrical streamers soared.

Tesla also needed financing, and he set his sights on John Jacob Astor IV, a member of Teddy Roosevelt's Rough Riders and a great-grandson of the wealthy fur trader and real-estate investor. "Jack" Astor was a likely target because his assets of some ninety million dollars were three times those of J. P. Morgan, and he appreciated inventors, noting his own patents for a bicycle brake, pneumatic walkway, and storage battery. Like Tesla, Astor also enjoyed prophecies, as evidenced by the science fiction novel he wrote in 1894, at the age of thirty, about space travel, entitled *A Journey in Other Worlds*. Yet Astor was known as an aimless dilettante, and not a nice one at that. A biographer described him as "cold-hearted, humorless, weak-minded and almost completely absent of personality."[1]

Tesla proved to be a better inventor than fundraiser. During a couple of meetings in New York City, Astor made it quite clear any investment he might make should help develop and market a new and profitable lighting system, yet the scientist persistently asked that the funds be used for his wireless power experiments. In addition to not respecting his investor's wishes, Tesla undiplomatically told Astor: "I am often and violently attacked because my inventions threaten a number of established industries," a statement that probably did not endear him to a cautious investor in those enterprises. Tesla even informed this militaristic Rough Rider that his inventions would "render large guns entirely useless, and will make impossible the building of large battleships."[2]

Astor, not surprisingly, would have none of it, at least initially. "You are taking too many leaps for me," complained the investor. "Let us stick to oscillators and cold lights [the precursor of florescent lamps]. Let me see some success in the marketplace with these two enterprises, before you go off saving the world with an invention of an entirely different order, and then I will commit more than my good wishes."[3]

Not discouraged, Tesla responded with blandishment, calling Astor "a prince among wealthy men, a patriot ready to risk his life for his country, a man who means every word he says." Such flattery prompted the egotistical investor to begin negotiating. "Your letter offering me some of your oscillator stock [has been] received," he wrote. "95 [dollars per share] seems rather a high price; for though the inventions covered by the stock will doubtless bring about great changes, they may not pay for some time as yet, and, of course, there are always a good many risks."[4]

After the scientist lowered his price, promised to focus on fluorescent lights, and obtained small risk-reducing contributions from George Westinghouse and Edward Dean Adams, Astor agreed to provide one hundred thousand dollars (equivalent to almost three million dollars today) for five hundred shares of Tesla Electric Company and a seat on the board of directors. (Tesla's subsequent diversion of funds to his wireless project would prompt Astor to stop payments after his first installment of thirty thousand dollars.)

With a site and financing, Tesla, then almost forty-three years old, made his move in May 1899. On his way by train to Colorado, the inventor stopped in Chicago to address the Commercial Club, composed of the city's thirty-two most powerful businessmen. Showing an "apparent difficulty in expressing his ideas," Tesla's "disjointed and disconnected" speech suggested that the inventor was losing some of his magic. Club members, reported one journalist, were "greatly mystified by the lecture and the experiments." Misreading his audience, who wanted to understand investment opportunities associated with electricity, Tesla began vaguely: "There is no one who does not speculate about the questions of his existence, asking whence he comes, whither he is going and what in reality he is."[5]

Having missed a golden opportunity to raise more money in Chicago, Tesla traveled on to Colorado Springs, where he moved into room 207 (divisible by three) of the Alta Vista Hotel, within walking distance of the new Colorado Springs lab. His windows looked out over the Rocky Mountains, and the hotel supplied him each morning with eighteen fresh towels. Town leaders welcomed the science celebrity with a grand banquet at the El Paso Club, where the inventor suggested he would be conducting wireless telegraphy experiments and trying to send signals from Pikes Peak to Paris. "I propose," he added, "to propagate electrical disturbances without wires."[6]

While Tesla moved to Colorado, George Scherff, his office manager, stayed in New York to supervise a team of craftsmen. Virtually each day, the two exchanged letters or telegrams, with the scientist frequently drawing diagrams for how he wanted equipment devised. Far away from friends and social activities, Tesla craved contact, asking Scherff to "write as often as possible"[7] and "make your correspondence more interesting."[8] He also issued his directives firmly: "All work must be done first class. Tell the workmen not to lose sight of this."[9] Despite imposing such pressure, Tesla inspired his staff, with Scherff commending "your wonderful work [in Colorado], and we know that, instead of a century, you are a thousand years ahead of the others."[10]

Most of the manager's notes responded to the scientist's instructions: "The sketches mentioned were not quite ready last night and could only be mailed this morning," Scherff wrote in one of the regular updates. "The work on the condenser is being pushed. Mr. Clark is milling down the teeth of the disk for the break.... Willy has finished grinding out the necks of the bottles and is now making salt solution."[11]

As would be expected with any long-distance business operation, some communications focused on finances. In this case, they demonstrated Tesla's tight resources. "As I have had to pay cash for everything," wrote Scherff, "the drain on the funds has been rather heavy," to the point where he needed an advance of two hundred dollars from the Chatham Bank in order to meet that week's payroll.[12] Some letters dealt with personnel issues, as when Tesla noted that one employee "has turned out very badly and I am compelled to send him away.... [We] were deeply abhorred to discover his character."[13] A few notes simply kept up on personal news, as when Scherff asked, "Permit me to inquire about your health. I hope that you may speedily recover from your cold."[14]

Away from the electrical interference within New York City's power and telephony systems that made it difficult to tune sensitive transmitters and receivers, Tesla set out to measure if the earth maintained an inherent electrical charge. If so, he reasoned, inserting a relatively small amount of power, and taking advantage of resonance, could transmit electricity and messages through the entire planet. Using a device called a coherer, which measures variations in energy forces, Tesla concluded that the earth was "literally alive with electrical vibrations."[15]

Colorado Springs, which adopted the nickname "Little London," also offered Tesla relief from snooping competitors. On the open plain, he built fences and placed signs around his property, declaring "Great Danger: Keep Out!" and he wryly inserted a quotation above the laboratory door from Dante's *Inferno*, warning: "Abandon all hope, ye who enter here." One enterprising reporter jumped the fence and ignored the signs in order to peer through the lab's window—only to confront Tesla's growling assistant: "Your life is in peril and you would be a great deal safer if you would remove yourself from the vicinity."[16]

Most mornings, Tesla worked at his office in the laboratory's southwestern corner, sending notes to Scherff, answering other correspondence, and preparing patent applications. The inventor and his on-site assistants, Kolman Czito and Fritz Lowenstein, would engage in endless discussions. Lowenstein, a twenty-five-year-old immigrant from Czechoslovakia, spent significant time with the inventor, eating lunches and dinners together, until September 1899, when, according to a biographer, a jealous Tesla became angry after discovering letters from his assistant's fiancée.[17]

Experiments would begin about three or four in the afternoon, and when the switch was thrown, enormous sparks and loud crashes would dance across the interior. Each evening, Tesla would record his day's notes, complete with detailed diagrams and mathematical formulas. (According to Aleksandar Marincic, who several decades later compiled approximately 375 pages of notes, the inventor devoted about 56 percent of his time in Colorado to the transmitter or high-frequency generator, about 21 percent to receivers for small signals, 16 percent to measuring the capacity of the vertical antenna, and the remaining 6 percent to other research.)[18]

One young part-time assistant, who swept the lab and ran errands, remembered Tesla as "tall and thin and nervous in manner." The inventor one day called this local lad into the office and asked how much he thought he was worth. The assistant was "tongue-tied with embarrassment," which Tesla supposedly watched "in amusement" and then asked if four dollars a week [about one hundred dollars today] would be about right. Since this was the young man's first job, he thought the sum was "princely," and became overwhelmed when the generous Tesla gave him eight dollars.[19] The inventor, however, could be very abrupt, as when he called that assistant "a little fool" when he made a minor mistake.

The Colorado plateau offered a stark environment and a dry and rarefied atmosphere. According to Tesla, "The sun's rays beat the objects with fierce intensity" and caused his tinfoil coatings to shrivel. Yet the "perfect purity of the air, the unequaled beauty of the sky, the imposing sight of a high mountain range, the quiet and restfulness of the place . . . [also] contributed to make the conditions for scientific observations ideal."[20]

The experiments, however, presented dangers. One day when assistant Czito had walked into town to pick up supplies, Tesla closed a special switch he had designed to handle heavy currents and went behind the coil to examine something. When the switch accidentally opened, "the whole room had filled with streamers, and I had no way of getting out." He tried breaking a window but could not locate any tools. With five hundred thousand volts breaking down the air and making it hard to breathe, he fell on his stomach and began to crawl. Sixteen-foot streamers sparked above him, yet he finally managed to shut the switch. Later he would note calmly, "The building [had] begun to burn. I grabbed a fire extinguisher and succeeded in smothering the fire."[21]

Tesla's experiments, according to one local writer, produced "the conviction that hell was breaking loose and belching into the building."[22] After Tesla fired up his equipment, a low rumble built into a "roar that was so strong it could be plainly heard ten miles away." The barn regularly emitted a ghostly blue light, sulphurous fumes of ozone filled the air, and the Tesla coils emitted fiery flashes, some more than a hundred feet in length.

Even the natural environment provided striking—and revealing—lightning discharges, "sometimes of inconceivable violence." On one occasion, noted Tesla, "approximately twelve thousand discharges occurred in two hours."[23] A particularly strong storm broke loose in the early evening of July 3, 1899, a day Tesla claimed he would never forget. The scientist's sensors recorded the fiery blasts hitting the earth, and after the storm passed, as expected, the indications became fainter and fainter. Yet those recordings soon returned and grew stronger, until they gradually decreased and ceased once more. The cycle continued, even though the storm had moved on some two hundred miles. To Tesla, this "wonderful phenomenon" proved the planet behaved like an electrical conductor that reflected the lightning's disturbances and created stationary waves.[24] He had worried the earth might act as a vast reservoir and the electrical waves would slowly dissipate, yet the evening's observations caused him to conclude the earth is "responsive to electrical vibrations of definite pitches, just as a tuning fork is responsive to certain waves of sound."[25] Tesla suggested the resonating electromagnetic charges acted "like an immense pendulum, storing indefinitely the energy of the primary exciting impulses" (i.e., the lightning strikes), and then he argued: "Not only was it practicable to send telegraphic messages to any distance without wires, as I recognized long ago, but also to impress upon the entire globe the faint modulations of the human voice, far more still, to transmit power, in unlimited amounts, to any terrestrial distance and almost without loss."[26] If stationary waves could be prompted by lightning, he reasoned, "it is now certain that they can be produced with an oscillator."[27]

The July 3 observations prompted Tesla to abandon his initial idea "to erect two terminal stations, one in London and one in New York," and to suspend great disks from a series of balloons that floated five thousand feet above the earth. He had planned to flash messages from an oscillator atop one of the towers to the first disk, which then would transmit it thousands of miles through the strata of rarefied air to the second disk, until the message reached the other tower. Unfortunately, Tesla's New York workshop initially could not apply enough varnish to stop leakages and achieve full inflation of the balloons, with one of the craftsmen complaining: "One of them has now twelve coats of varnish, but it is still impossible to fill it up with air. The varnish remains sticky and comes off in patches."[28] Even when Tesla finally launched balloons, they proved to be too heavy, being weighed down by the lengthy electrical wires attached to the ground station.

After abandoning the balloon approach, Tesla turned instead to directing his artificial lightning into the ground. Taking advantage of his previous observation of the earth's natural electrical vibrations, he sensed his earthbound discharges produced stationary waves—when two ripples of identical frequency but traveling in opposite directions interfere with one another. That remarkable measurement led him to declare, "Impossible as it seems, this planet, despite its vast extent, behaves like a conductor of limited dimensions,"[29] meaning a low-frequency stationary wave generated by his oscillator, similar to a lightning strike, could prompt enough resonance to send messages and power around the world. Tesla described his observation as "the great truth accidentally revealed and experimentally confirmed." This planet, he concluded, "with all its appalling immensity, is to electric currents virtually no more than a small metal ball."[30]

More specifically, Tesla theorized he could reinforce the electrical echo from his initial artificial lightning and, as a result of resonance, send it back through the earth even more powerfully. He suggested that anyone on the planet could plug into the ground and draw off some of this energy from the pulsing waves prompted by Tesla's oscillators in order to light lamps or run machines.

To describe this resonance, while expanding on his "sensitive trigger" observation made during a thunderstorm in Lika, Tesla offered a sports analogy, suggesting his power-thrusting plan was similar to a pump forcing air into an elastic football. "At each alternate stroke the ball would expand and contract," he explained. "But it is evident that such a ball, if filled with air, would, when suddenly expanded or contracted, vibrate at its own rate. Now if the strokes of the pump be so timed that they are in harmony with the individual vibrations of the ball, an intense vibration or surging will be obtained."[31] Similarly, the inventor planned to pump electrical energy at the right resonance into the earth in order to transmit power and messages around the world.

Tesla, however, needed a stronger pump, or, more accurately, a more powerful magnifying transmitter. In his New York lab, his equipment had reached four million volts and produced sixteen-foot discharges, but in Colorado he sought fifty million volts and hundred-foot-long sparks.

The inventor adjusted numerous variables. He changed the size of the capacitors that fed the primary coil and tried several different coils for the secondary. He tested extra coils placed in different places around the transmitter's circuit. He raised the height of the spherical terminal atop the station's roof. He increased the voltage coming into the barn from the Westinghouse transformer. He also tried different tunings for each of the coils in order to enhance resonance. The result was longer and longer streamers that jumped ever-expanding gaps between his coils.

Tesla's most dramatic demonstration of his new power occurred on a clear autumn evening when the confident inventor claimed he would obtain "the first decisive experimental evidence of a truth of overwhelming importance for the advancement of humanity."[32] Dressed elegantly in a black Prince Albert coat, gloves, and a dark derby hat, Tesla discussed a plan with his brave assistant, Kolman Czito, to explode millions of volts from the top of his mast. Czito would be in charge of the switch, while Tesla went outside to view the copper dome.

Tesla commanded: "When I give you the signal, I want you to close the switch and leave it closed until I give you the signal to open it." Czito complied and as the heavy current surged, snakes of flames writhed up the mast. A blue aura, dancing sparks, vicious snaps, and the sharp smell of ozone filled the laboratory. Although Czito couldn't see the scientist standing outside, a blissful Tesla stared at the man-made bolts shooting an estimated 135 feet from the top of the mast. The inventor had created his own lightning, something previously reserved for Mother Nature.

Suddenly, and with no warning, the roar and blazes ceased. All became dark. An angry Tesla yelled to Czito: "Why did you do that?" Yet the assistant pointed at the still closed switch.

Tesla then called the El Paso Power Company, screaming: "This is Nikola Tesla. You have cut off my power! You must give me back power immediately! You must not cut off my power!"

The town's distraught power manager yelled back, "You've thrown a short circuit on our line with your blankety-blank-blank experiments and wrecked our station. You've knocked our generator off the line, and she's now on fire."[33]

Tesla's artificial lightning, while it lasted, had produced strange effects on the planet's surface. The energized ground around the laboratory, for instance, snapped as he walked across it; he was protected by shoes with six inches of insulating cork attached to the soles. Scores of animated butterflies swirled hypnotically in tight circles above the charged earth. Horses with iron shoes in nearby fields became agitated and reared up on their hind legs. Independent lightning rods on neighboring buildings flashed when Tesla's coil discharged, and water faucets emitted sparks. Perhaps most significant, three bulbs placed atop a square of wire sixty feet from the experimental station glowed.

Tesla compared the phenomenon of the earth's conductivity to the echo heard when the sound of a voice is reflected off a distant wall. In Colorado Springs, he felt he reflected an energy wave from within the earth and produced "an electrical phenomenon known as a 'stationary' wave—that is, a wave with fixed nodal and ventral regions." Such waves in the earth, he concluded, would enable an array of wireless effects. According to Tesla, "By their use we may produce at will, from a sending station, an electrical effect in any particular region of the globe; we may determine the relative position or course of a moving object, such as a vessel at sea, the distance traversed by the same, or its speed; or we may send over the earth a wave of electricity traveling at any rate we desire, from the pace of a turtle up to lightning speed."[34]

These insights convinced the scientist his wireless system would have far greater reach than Marconi's apparatus, which in March 1899 successfully sent a message across the English Channel from France to England. Although Tesla never provided a public demonstration, he boasted to the *New York Journal* that he would soon be able to transmit private, wireless communications around the entire world. "You will be able to send a 2,000 word dispatch," he declared, "from New York to London, Paris, Vienna, Constantinople, Bombay, Singapore, Tokyo, or Manila in less time than it takes now to ring up 'central.'"[35]

Tesla later would claim the title of "Father of the Wireless" because his radio efforts had begun eight years before and had achieved "a number of radical improvements." He appreciated early on that his first requirement was to develop high-frequency generators and electrical oscillators. For the next step, he said, "the energy of these had to be transformed into effective transmitters and collected at a distance in proper receivers." Tesla, in fact, seems to have been the first to propose using transmitting and receiving antennas tuned to the same frequency. Also needed were systems to ensure each transmission could be kept separate and free of extraneous interference. The scientist expressed great pride in his advances and declared boldly: "I gave to the world a wireless system of potentialities far beyond anything before conceived." He went on to brag he had "overcome all obstacles which seemed in the beginning unsurmountable and found elegant solutions of all the problems which confronted me."[36]

Tesla argued that no one, including Marconi, could transmit a message any appreciable distance "without the use of my apparatus." He claimed five of his discoveries were necessary for such transmission: "First, my method of oscillatory conversion by means of condensers; second, the so-called 'Tesla transformer'; third, my apparatus for the transmission of energy without wire, comprising grounded, resonant circuits; fourth, my methods and apparatus for individualizing signals; and, fifth, my discovery of the stationary waves."[37]

Tesla had filed his basic radio patent application in the United States in 1897, and it was granted three years later. He had proposed a wireless "system of transmission of electrical energy" that would "transmit intelligible messages to great distances."[38] (Tesla submitted other radio patents, focused on transmissions through the earth, after his Colorado Springs experiments.) Although Marconi had applied in England in 1896, he did not file in America until November 1900, and he was turned down. The Italian inventor resubmitted his claim over the next three years, consistently arguing he was unaware of Tesla's ideas or devices. The U.S. Patent Office kept rejecting the pleas and in 1903 bluntly concluded: "Marconi's pretended ignorance of the nature of a 'Tesla oscillator' being little short of absurd. The term 'Tesla oscillator' has become a household word on both continents [Europe and North America]."[39] A Marconi associate later admitted the Italian also did not understand the earth's role in conveying electrical energy; "we knew nothing," he stated, "about the effect of the length of a wave transmitted governing the distance over which communication could be affected."[40]

Still, the Marconi Wireless Telegraph Company—in part because of the young Italian's dramatic demonstrations, his business acumen, and his family connections with wealthy English aristocrats—thrived on stock markets, with British shares rising quickly from three dollars to twenty-two dollars. In the United States, the inventor attracted investments from Andrew Carnegie and engineering advice from Thomas Edison. For reasons never explained, but probably because of pressure from such financial and scientific backers, the U.S. Patent Office retreated from its previous rejections and in 1904 granted a radio patent to Marconi.

Tesla tried to remain stoic as the Italian gained more and more attention. When an engineer said, "Looks as if Marconi got the jump on you," Tesla replied, "Marconi is a good fellow. Let him continue. He is using seventeen of my patents."[41] In fact, Tesla early in his wireless-development process graciously met with the Italian scientist to explain "the function of my transformer for transmission of power to great distances." On other occasions, however, Tesla bluntly asserted his system's superiority and argued, "Mr. Marconi was experimenting with my apparatus unsuccessfully." He once called the Italian scientist "a donkey."

As Marconi became more famous, it seems he avoided Tesla. When the Italian scientist traveled to New York to raise money and license his own patent, Tesla tried to confront Marconi at one of his presentations but wrote: "When he learned that I was present he became sick, postponed the lecture, and up to the present time has not delivered it."[42]

Tesla's system initially offered more privacy and noninterference of messages. When Marconi in 1901 tried to cover the America's Cup race off New York City, the American Wireless Telephone & Telegraph Company, a rather aggressive competitor, sent ten-second transmissions that jammed the Italian scientist's signals. Marconi's radio broadcasts also could be picked up by other receivers, a real drawback for the

U.S. military worried about the enemy intercepting messages. In contrast, Tesla drew on the "individualization" advances he made during his robotic boat experiments. He claimed his transmitter "is not primitively characterized by a single note, or peculiarity . . . but represents a very complex and, therefore, unmistakable individuality, of which the receiver is the exact counterpart, and only as such can it respond."[43]

Even though Tesla enjoyed a reasonable claim to being the first to theorize and develop wireless communications equipment, Marconi was winning the public relations battle and demonstrating actual transmissions. More and more reporters heard Tesla making promises while Marconi achieved progress. The gossipy *Town Topics* went so far as to disdainfully describe Tesla as "America's Own and Only Non-Inventing Inventor, the Scientist of the Delmonico Café."[44]

What was happening to Tesla? No doubt he was increasingly isolated. Encamped in Colorado, he failed to network with his fellow inventors and trade-journal writers who became more and more skeptical when his prophesying never materialized. The laboratory fire in March 1895 certainly set back Tesla's efforts and allowed others, particularly Marconi, to make unchallenged progress, and it also diverted his focus away from realizing his wireless transmission ideal. Tesla's lectures seemed to suggest he was stuck on displays of artificial lightning, producing brighter snaps and louder crackles rather than practical applications.

The public skepticism increased after Tesla's highly sensitive receiver in Colorado Springs detected weak oscillations in the form of rhythmic beeps, which the scientist suspected came from outer space. "My first observations positively terrified me," wrote Tesla later, calling the sounds "something mysterious, not to say supernatural." As he thought about the phenomenon, he rejected any possible solar or terrestrial cause and admitted "the feeling is constantly growing on me that I have been the first to hear the greeting of one planet to another."[45]

Tesla initially kept such observations to himself. Yet when asked by the local Red Cross Society to opine on the "greatest possible achievements of the next hundred years," he could not hold in his excitement.

"I have observed electrical actions, which have appeared inexplicable," he gushed. "Faint and uncertain though they were, they have given me a deep conviction and foreknowledge, that ere long all human beings on this globe, as one, will turn [their] eyes to the firmament above, with the feelings of love and reverence, thrilled by the glad news: 'Brethren! We have a message from another world, unknown and remote. It reads: one . . . two . . . three.'"[46]

Although Tesla merely suggested the signals were extraterrestrial, scornful journals ridiculed him for having conversations with Martians, and scientists scoffed. The former director of the University of California's Lick Observatory complained, "Until Mr. Tesla has shown his apparatus to other experimenters and convinced them as well as himself, it may safely be taken for granted that his signals do not come from Mars."[47]

Tesla, of course, was not the only scientist who claimed to hear communiqués from outer space or the beyond. Marconi almost twenty-five years later said he "had often received strong signals out of the ether which seemed to come from some place outside the earth and which might conceivably have proceeded from the stars."[48] Edison made an even more bizarre claim of having devised a "spirit phone" that could contact the dead.[49] The editor Robert Johnson, trying to protect his friend from criticism, concluded that the moral was: "Better to be second than first in some things."[50]

Notwithstanding such diversions and criticisms, Tesla achieved some successes during his eight months in Colorado Springs. He reached his goal of fifty million volts from a magnifying transmitter, an advanced version of the Tesla coil, by carefully tuning and synchronizing his equipment. He created artificial lightning, with thunder heard fifteen miles away in Cripple Creek and streamers stretching 135 feet. After fixing the El Paso Power Company's dynamo, he manipulated different-sized streamers flying off the sphere rising above the roof of his barn. Tesla certainly enjoyed the show, commenting that "at night, this antenna, when I turned on to the full current, was illuminated all over and it was a marvelous sight."[51] More importantly, he demonstrated the power of high-voltage transmissions, yet he only suggested their potential to transmit electricity and messages.

Tesla sometimes complained in Colorado that his "concentration of mind strained to so dangerous a degree the finest fibers of my brain."[52] Yet he regularly conveyed his enthusiasm to friends in New York. "My experiments here have been a source of great pleasure and satisfaction," he wrote, "and if I never came here I would never had discovered facts which I did."[53]

The most striking results from Colorado Springs proved to be the series of sixty-eight pictures taken at the experimental station by Dickenson Alley, a photographer sent from New York by *The Century Magazine*. Among the more dramatic were of the lengthy streamers emitted by Tesla's transmittor. Perhaps the most memorable image was of Tesla calmly reading a book in front of a deluge of sparks; yet this photograph proved little about the inventor's experiments. Alley had cleverly used a double exposure since it would have been too dangerous for the scientist to actually sit so close to such an electrical storm. Tesla later admitted, "The streamers were first impressed upon the plate in dark or feeble light, then the experimenter placed himself on the chair and an exposure to arc light was made and, finally, to bring out the features and other detail, a small flash powder was set off."[54] The publicity-conscious experimenter claimed he didn't like such trick photography, but he argued, unconvincingly, that "some people find such photographs interesting."[55]

Tesla also asserted that the high-frequency current he sent into the ground lit two hundred 50-watt incandescent lights in the mountains twenty-six miles from Colorado Springs, but no evidence supports that assertion. An Alley photograph, however, does show three glowing lamps—placed on the ground atop a sixty-two-foot square network of wire—that were said to be sixty feet from the transmitter in the laboratory. (*The Prestige*, a 2006 movie starring Hugh Jackman and Christian Bale, exaggerated the Tesla claim with a scene complete with a fog-drenched field and soaring music, which showed the inventor walking amidst a hundred lighted bulbs he had screwed into the ground.)

Why would Tesla exaggerate? One explanation is that he tried to compensate for not achieving his one grand ideal—the wireless transmission of power. Two hundred bulbs twenty-six miles away over the Rocky Mountains must have sounded far more dramatic to this consummate showman than only three bulbs a mere sixty feet away. Yet the false claim suggests that Tesla, for perhaps the first time, was willing to go beyond showmanship to embellish the truth.

The inventor maintained that the photographs effectively documented his work, saying "nothing could convey a better idea of the tremendous activity of this apparatus."[56] Unfortunately, he had no independent witnesses to his wireless transmission tests, something out of character with his earlier experimentations.[57] Even Tesla's chief assistant admitted during a subsequent patent trial that he had remained in the lab and had never personally seen man-made lightning during the high-frequency experiments. That lack of verification prompted critics, as well as investors, to increasingly question the value of the inventor's wireless transmission claims.

This dearth of independent verification is odd for a careful scientist like Tesla. Why would he feel that his notes and photographs were sufficient to impress financiers and other scientists? Was he so ensconced in his own little world in Colorado Springs that he simply got sloppy? Was he so addicted to his own pyrotechnics that he abandoned the practical? It seems Tesla was losing his rigor, his focus. He was not, however, losing his self-confidence.

Tesla expressed complete satisfaction with his Colorado achievements, arguing his aim had been simply "to perfect the apparatus and make general observations."[58] He had created lightning and pushed power into the earth. He even believed his large oscillator had set the earth in electrical resonance, opening up the possibility for people around the world to place receptors into the ground in order to obtain signals and power. As the twentieth century arrived, still committed to his one big idea, Tesla wanted to move on to even higher frequencies and unlimited wireless transmissions.

11.
SHEER AUDACITY
Wardenclyffe, Long Island

esla returned to New York City in January 1900 with a bold plan to build a giant transmission tower and a profitable business around the electrical phenomena he had observed in Colorado Springs. Confident of his prospects, he upgraded to the luxurious Waldorf Astoria, the world's tallest hotel and the first to provide complete electric service and private bathrooms. Located on Fifth Avenue and 34th Street, the hotel attracted many of the world's wealthiest and most celebrated, including Andrew Carnegie, Prince Henry of Prussia, and Arctic explorer Frederick Cook. As part of his investment deal with John Jacob Astor, who was one of the hotel's owners, Tesla would live there for the next two decades.

Tesla filed three patents that year for wireless communication devices and bragged to the media that "my experiments have been most successful, and I am now convinced that I shall be able to communicate by means of wireless telegraphy not only with Paris during the [upcoming] Exhibition [Universelle of 1900], but in a very short time with every city of the world."[1]

A key part of the inventor's public relations blitz featured his article and Dickenson Alley's photographs from Colorado Springs for *The Century Magazine*, that well-respected venue edited by his good friend

Robert Johnson. Tesla's piece was to be in the same issue as an article on political reform by Theodore Roosevelt, then nominated to be vice president. In front of this wide audience, Johnson hoped Tesla would explain the historical importance of his discoveries; instead, the inventor used this platform to pontificate.

Tesla's first draft highlighted lifestyle advice. "Whisky, wine, tea, coffee, tobacco, and other such stimulants are responsible for the shortening of the lives of many, and ought to be used in moderation," he wrote. "But I do not think that rigorous measures of suppression of habits followed through many generations are commendable. It is wiser to preach moderation than abstinence."[2] In the same article, however, he admitted that he'd consumed enough alcoholic beverages to "form a lake of no mean dimensions."[3]

Tesla even offered recommendations about hygiene: "Uncleanliness, which breeds disease and death, is not only a self-destructive but a highly immoral habit. In keeping our bodies free from infection, healthful and pure, we're expressing our reverence for the high principle with which they are endowed. He who follows the precepts of hygiene in this spirit is proving himself, so far, truly religious."[4]

Why would Tesla offer lifestyle and hygiene advice? In his mind, the suggestions had something to do with inventing. The Scientific Man, as he envisioned that ideal, needed to be moderate and clean.

Of course, that model inventor also had to offer broad thinking, and Tesla proposed later in the article a global system of wireless communications. He predicted controlling the weather with electrical energy. He foresaw a world in which machines eliminated the need for war.

The ever-expanding article became a source of conflict with Johnson, who complained that Tesla's ramblings offered more visions than facts. "Trust me in my knowledge of what the public is eager to hear from you," the seasoned editor wrote. "Keep your philosophy for a philosophical treatise and give us something practical about the experiments themselves."[5]

Tesla responded testily: "My opinion is that no extraordinary intelligence, such as possessed by every *Century* reader, is needed to follow my ideas and to recognize in them a connected whole." With sarcasm, he asked, "How then does the *Century* differ from other magazines?"[6]

Ignoring Johnson's advice, Tesla provided thirty-six pages of florid prose and observations about human progress, including an incomprehensible formula: $E = mV^2/2$—where "E" equaled total human energy, "m" the mass of humanity, and "V" the velocity of human change. He even concluded with a poem entitled "Hope" from Johann Wolfgang von Goethe: "Lo! These trees, but bare poles seeming, yet will yield both fruit and shelter."

Johnson protested again. "I will do anything but see you write an ineffective article when you have a chance to make so good an impression." The editor offered numerous suggestions and complained the writer was "making a task of a simple thing." He also pressed for quick action: "Give us something practical . . . and do it soon." He even tried flattering Tesla: "Forgive my clumsy way of saying it because of my love and respect for you and because I have had nearly thirty years of judging what the public finds interesting."[7]

Tesla answered with a mix of affection and humor: "I learned with great regret just now, that you are not feeling well, and I shall cherish the two hopes, that you will soon get well and that it is not my article which made you sick."[8] Tesla demanded keeping the complex formula and Goethe poem in his article, but he tried to appease Johnson by agreeing "that to preserve unity, I must have only three chapters after the first introductory part. Each of the chapters will thus deal with a specific problem and its solution, and they ought to be, to preserve balance, of not widely different lengths. You will recognize the great advantage of this mode of treatment."[9] Johnson described the modified text to be "a vast improvement."[10]

Tesla titled the *Century* article "The Problem of Increasing Human Energy, with Special References to the Harnessing of the Sun's Energy." He paid particular attention to energy efficiency, or eliminating waste in the generation and use of electricity; he calculated that we "do not utilize more than two per cent of [coal's] energy." The inventor, probably thinking of himself, declared: "The man who should stop this senseless waste would be a great benefactor of humanity." Yet Tesla recognized that even burning coal more efficiently would "ultimately lead to the exhaustion of the store of the material." Needed, instead, was "to turn to the uses of man more of the sun's energy." He estimated those inexhaustible or renewable rays "supply energy at a maximum rate of over four million horsepower per square mile." [11]

Although he discussed at length the potential to generate electricity from the sun, wind, and geothermal resources, the scientist came to the conclusion that "the 'solar engine,' a few instances excepted, could not be industrially exploited with success." [12] Yet he offered a grander sustainable-energy proposal, one that would utilize the temperature differences between the earth's internal heat and the cold of the cosmos. "Imagine a thermopile consisting of a number of bars of metal extending from the earth to the outer space beyond the atmosphere," he wrote. "The heat from below, conducted upward along these metal bars, would cool the earth . . . and the result, as is well known, would be an electric current circulating in these bars." [13]

Near the end of his article, Tesla shared his vision of the inventor's role—to focus on long-term fundamental ideals rather than short-term practical results. "The scientific man," he wrote, "does not aim at an immediate result. He does not expect that his advanced ideas will be readily taken up. His work is like that of a planter—for the future. His duty is to lay the foundation for those who are to come, and point the way." [14]

The *Century* article attracted a lot of attention in the popular press, reflecting Tesla's growing reputation in both the United States and Europe as a scientific genius. Yet engineering journals tended to be less kind, with one suggesting "the public should be protected from such wild speculation passing for scientific fact." [15]

Skepticism was justified. Tesla had been given the opportunity to address a wide audience, but he couldn't stay on message, and he focused far more on visions than practicalities. One explanation is his self-confidence had grown so much that he overestimated his readers' interest and the power of his own pontificating. It also could be said that he was losing his filter, with his mind jumping wildly from hygiene to geothermal energy to wireless transmissions.

The article, however, hinted at the next phase of Tesla's life and work. In flowery terms, he resolved to develop "an efficient apparatus for the production of powerful electrical oscillations." Such a "venturesome task," he declared, involved "great sacrifice," yet it was "the key to the solution of other most important electrical and, in fact, human problems."[16] Tesla, in a subsequent interview, argued that the insights he obtained in the Rocky Mountains could be realized only if he built a larger and more powerful system. "The plant in Colorado," he explained, "was merely designed in the same sense as a naval constructor designs first a small model to ascertain all the quantities before he embarks on the construction of a big vessel."[17]

Building something big, however, required big investments. While Astor had abandoned Tesla as a fickle scientist, he allowed him to remain rent-free in the Waldorf Astoria. Tesla felt his best hope was George Westinghouse, with whom he had developed a personal relationship and whom he called "my friend," even though he addressed his letters formally to "Dear Mr. Westinghouse." The two exchanged regular notes, with Tesla once suggesting a portrait painter with "a deep idea of your personality such as would enable him to make of you a portrait which would be a very valuable addition to the art."[18] Westinghouse was all business. On several occasions, the Pittsburgh manufacturer asked about a "device for commutating alternating into direct current:... I would like to know if you have advanced far enough with your patents and experiments to be now in a position to let the Westinghouse Electric & Mfg. Co. undertake the manufacture of such apparatus upon reasonable terms."[19]

Tesla wanted to talk about little else than his wireless experiments. Not long after returning from Colorado Springs, he wrote a three-page "personal" letter to Westinghouse proclaiming that his "success has been even greater than I anticipated, and . . . I have absolutely demonstrated the practicability of the establishment of telegraphic communication to any point of the globe by the help of a machine I have perfected." Tesla then bemoaned that he hadn't secured sufficient funding beforehand and inadvertently revealed his lack of business skills: "I have been so enthused over the results achieved and have worked with such a passion, that I have neglected to ask such provisions for money, as would have been dictated by prudence." He concluded by admitting, "I make no secrecy of needing the money." [20]

The different motivations of Tesla and Westinghouse were on display after a court ruled in their favor on a patent lawsuit regarding Tesla's alternating-current motor. Westinghouse argued Tesla sought recognition while he wanted profits: "The courts have awarded to you the credit for a great invention; and because we are likely to have our rights respected [the Westinghouse company] may reap some of the advantages due to the large expenditures we have made in the commercial development of your inventions." [21]

Still, Westinghouse didn't see the business potential in Tesla's latest wireless research and refused to support him, although his company did offer a loan on the condition the scientist used Westinghouse equipment. Tesla kept pleading, to little avail. "Has anything happened to mar the cordiality of our relations?" he asked earnestly. He ardently argued that wireless transmissions would "very soon create an industrial revolution such as the world has never seen before." With that would come the possibility for great profits, and he appealed to Westinghouse's instincts as an inventor: "This far advanced work [is] now in the delicate stage of a sprout. It is easy to destroy and difficult to succeed." [22]

Tesla could have made money by working with the United States Lighthouse Board, which wanted the scientist in spring 1899 "to establish a system of wireless telegraphy upon the Light-Vessel No. 66, Nantucket Shoals, Mass., which lies off about 60 miles south of Nantucket Island." [23] The government "preferred to award home talent"—meaning Tesla over Marconi—but the proud American citizen disliked being compared in any way to the Italian scientist and claimed he "would never accept a preference on any ground." [24] Such stubbornness meant Tesla missed the chance to develop wireless communication on someone else's dime. According to one historian, the inventor's refusal became "one of the most significant blunders of his career," since the project would have brought in needed funds as well as offered high-profile publicity for Tesla's sophisticated long-distance wireless telegraphy.[25]

Thinking on a grander, if less practical, scale, Tesla proposed a "world system" of wireless communications. He envisioned sending telephone messages, music, stock market reports, pictures, and even secure military messages to any part of the planet. The man even foresaw the Internet access that did not become available, at least in the form of the World Wide Web, until 1990 and that we now take for granted. "When wireless is fully applied," Tesla predicted, "the earth will be converted into a huge brain, capable of response in every one of its parts."[26]

With that vision, Tesla eventually obtained a loan of $150,000 (equivalent to about $4.3 million today) from J. P. Morgan. Wall Street's most powerful financier was in the midst of forming a billion-dollar steel industry trust, and the amount he loaned Tesla was equal to what he'd paid for a single painting, *The Duchess of Devonshire* by Thomas Gainsborough. In supporting Tesla's devotion to wireless research, Morgan could have been trying to protect his investments in existing telegraph, telephone, and electricity businesses, including the General Electric Company. Or he could have been trying to gain control of any future wireless industry. He also could have simply wanted a way to signal his steamships at sea or to obtain up-to-date quotes from the New York Stock Exchange when he traveled in Europe. Perhaps Morgan considered the contribution "philanthropic," as Tesla hinted in a letter,[27] similar to his charitable support for the Metropolitan Museum of Art and other artists and intellectuals.

Whatever the motivation, Morgan cut a hard bargain, shamelessly adding demands to obtain control of Tesla's previous lighting patents. The scientist complained modestly to one of Morgan's assistants, writing: "I need scarcely say that I would sign any document approved by Mr. Morgan, but believe that there exists a misunderstanding in regards to my system of lighting which was not in the original proposition."[28]

As when Tesla tore up his royalty contract with George Westinghouse, here the inventor again does the irrational—giving away control of key and unrelated patents. Was he simply a poor negotiator? Because he was low on money, the Astor funds having run out, was he vulnerable to being tricked by a business superman? Was he doubting his own abilities and worrying he had no other options? Did he simply abhor the world of finance and want to get back to inventing? Some combination of such charges offers a perspective on Nikola Tesla in the early twentieth century.

Morgan usually ended up controlling the conglomerates he created, a process sometimes referred to as "Morganization." The investor, said Tesla, towered "above all the Wall Street people like Samson over the Philistines."[29] The man was imposing, standing a stocky six feet two inches tall with broad shoulders and piercing eyes. He suffered from a rare skin disease that caused his nose to grow huge and glow purple (the vain banker avoided pictures and required portraits be touched up), yet he flaunted his Don Juan conquests. The arrogant and feared banker cared little for his employees, competitors, the public, or even the inventors he bankrolled.

Tesla had little choice but to express appreciation for the "great man's" loan. "How can I begin to thank you in the name of my profession and my own, great generous man!" he wrote gushingly after signing the financing agreement. "My work will proclaim loudly your name to the world. You will soon see that not only am I capable of appreciating deeply the nobility of your action, but also of making your primarily philanthropic investment worth a hundred times the sum you have put at my disposal in such a magnanimous and princely way!"[30]

Tesla also tried to endear himself to the Morgan family. Louisa, J.P.'s daughter, invited the scientist to her wedding in the autumn of 1900, an event also attended by John Jacob Astor and Theodore Roosevelt. Anne, J.P.'s younger daughter, took a fancy to Tesla at that party and invited him to the grand Thanksgiving dinner at the Morgan mansion on Madison Avenue, where Tesla delighted the gathered with displays of streamers, wands of light, and various wireless robots.

Another fundraising success occurred in August 1901 when Tesla received from James S. Warden, a lawyer and real estate investor, a low-cost mortgage for a two-hundred-acre parcel of wooded and isolated land on the north shore of Long Island. Warden had purchased 1,600 acres of farmland in Suffolk County, sixty-five miles from New York City along the recent extension of the Long Island Railroad's Northern Branch, where he hoped to entice New Yorkers to build summer retreats. The plot, mostly surrounded by potato farms, was just a few miles from a charming beach at the hamlet of Wading River on Long Island Sound and not far from the town of Southampton on the Atlantic Ocean side. Warden christened the development "Wardenclyffe" (although the village's name changed in 1906 to "Shoreham"). No doubt the developer admired Tesla's brilliance but he also bet that the scientist's new project might attract other businesses and employ hundreds, if not thousands, of workers who would need new homes nearby.

Tesla convinced Stanford White—the famed and flamboyant architect he had met at The Players, the historic private social club facing New York's Gramercy Park—to design an imposing laboratory, a ninety-six-foot-square, one-story, red-brick building with a center chimney topped with a cast-iron wellhead. A principal in the influential design firm of McKim, Mead, and White, the architect, described in social columns as "the dashing man with the red moustache," had won praise for his work on the Washington Square Arch (1889), the second Madison Square Garden (1890), the Vanderbilt family residence (1905), the New York Herald Building (1894), as well as the "cathedral of power" at Niagara Falls. During the design and construction of the Wardenclyffe laboratory, White and Tesla became good friends, enjoying long walks together through Long Island's woods. White would drive up from his Long Island estate in his fashionable electric two-seater runabout, and Tesla sometimes would spend nights at the White compound, where, according to White's son, Tesla "used to wander around at night in the garden in the moonlight; and when my mother asked him why he wasn't asleep, he replied: 'I never sleep.' "[31] It was an odd combination—an inventor who abhorred physical contact and an architect who sponsored "Girl-in-the-Pie Banquets" that featured topless young women.

Tesla bet big on Wardenclyffe. The laboratory and tower culminated his drive toward wireless transmissions, his one big idea.

White initially estimated the project's construction expenses at fourteen thousand dollars, but they continued to rise, in part because the stock market panic of 1901 pushed up the cost of building materials. Oblivious to such increases, Tesla demanded enhancements, including handsome woodwork and a wide balcony. The structure, he said, was to be "divided into four compartments, with an office and a machine shop and two very large areas." In one of those large areas sat two 300-horsepower boilers. In the opposite room, there were "five large tanks, four of which contained special transformers created so as to transform the energy for the plant."[32] Tesla also installed a control unit that could "give every imaginable regulation that I wanted in my measurements and control of energy,"[33] and he arranged for the Long Island Railroad to build a spur that could deliver equipment and coal.

The scientist spared little expense personally as well. During the lab's construction, he commuted by train daily from the Waldorf Astoria to Wardenclyffe, accompanied by a man servant carrying a heavy hamper with lunch prepared by the fancy hotel's chefs. The hour-and-a-half trip aboard a luxury train car—past Manhasset, Oyster Bay (where Vice President Theodore Roosevelt lived), and Smithtown (where Stanford White had his family estate)—gave Tesla a chance to review mail and articles. When the Wardenclyffe facility became operational, he rented a lavish cottage on the shore of Long Island Sound.

Tesla, moreover, spent heavily on laboratory equipment and personnel. He employed nearly twenty highly skilled workmen, as well as armed guards to prevent competitors from spying.

Tesla in the early years of the twentieth century showed every sign of success—with an apartment at the Waldorf, funding from the greatest financier, and designs from the foremost architect. Despite hints of trouble associated with the lack of verifications for his Colorado Springs experiments and his ramblings in the pages of *The Century*, most reporters and fellow scientists predicted more and more achievements—

particularly because the brilliant inventor, with an impressive track record, was putting all of his energies into a single spectacular idea.

The tower at Wardenclyffe, Tesla argued, would have to be massive in order to successfully transmit signals and power through the earth. Wanting to use low-frequency long waves, he initially calculated the assembly should rise six hundred feet, about twice the height of the iconic Flatiron Building that opened near Madison Square in 1903 (and more than three times taller than many of today's cell phone towers). Noting that a sphere stored more electrical charge than other profiles, he called for topping the tower with a molded set of hemispheres. However, the ever-rising costs were now estimated by White at four hundred fifty thousand dollars—more than thirty times his original estimate. Tesla reluctantly accepted redesigns and settled on a 187-foot-tall tower topped with a simple dome, 68 feet in diameter.

The escalation of Wardenclyffe's costs from fourteen thousand dollars to four hundred fifty thousand dollars is an apt metaphor for Tesla's grandiosity, deserved or not. It demonstrates both his total commitment to this wireless ideal as well as his lack of business sense or even of moderation when it came to money. This brilliant man—born between one day and the next, between the present and the future—usually had a foot in both this world and the world only envisioned. Yet at Wardenclyffe, Tesla veered untethered toward the future. With little anchor to reality, let alone anyone to check his impulses, the inventor lost his way.

Although smaller than Tesla had hoped, the cupola still weighed fifty-five tons and could act like a sail in the strong winds from Long Island Sound and the entire tower might topple. To obtain the needed support, White designed an ingenious eight-sided structure that tapered as it rose. To ensure it didn't transmit electric shocks, Tesla ordered unfinished pine timbers rather than conductive iron or steel girders. For the system to transmit power through the ground, the inventor declared that the tower had "to have a grip on the earth so that the whole of this globe can quiver."[34] To gain such a grip, Tesla sank a shaft below the water table, constructed a 120-foot spiral staircase around it, and then drilled sixteen iron pipes an additional 300 feet into hard rock. The site's designs also called for a series of four tunnels rising to the surface from the bottom of the well, yet their actual location and purpose remain a mystery. Local villagers observed the enigmatic excavation was "as deep as the tower is high, with walls of masonwork and . . . that Tesla spends as much time in the underground passages as he does on the tower or in the handsome laboratory."[35]

Even though the inventor had trimmed Wardenclyffe's size, it was audacious in November 1901 to begin construction of such an expensive project when his finances were so uncertain. It also suggested imprudence when Tesla's chief supporter, a tycoon capable of making and breaking corporations, let alone dreamy inventors, showed displeasure in Tesla's idealism. Maybe the ever-confident scientist believed his tower would indeed launch a world communications enterprise. Maybe he remembered how his previous inventions had provided riches for Morgan and other financiers (if not for himself directly). Maybe his myopic focus on this one grand idea of wireless transmissions was simply foolhardy.

Undaunted, Tesla simply maintained that this "grip of the earth" would allow his magnifying transmitter to send an electrical wave through the earth and have it bounce back. Building on the theory of resonance, he assumed he could transmit another charge that would be in phase with the original, thus creating a stationary wave. With the earth humming at this electromagnetic resonant frequency and with his tower offering a return for the circuit, the scientist believed anyone could connect a receiver into the ground and obtain transmissions of messages and power. Expressing total confidence in his design, he assured Morgan the Wardenclyffe project would produce "electrical effects of virtually unlimited power, not obtainable in any other ways heretofore known."[36] Tesla even began making plans for a series of similar towers around the world, the second to be placed outside Glasgow, Scotland.

Yet Tesla faced growing competition from Marconi, who by early 1900 was sending messages up to 185 miles and who was building large circular antennas in Cornwall in England and St. John's in Newfoundland in order to transmit radio waves across the Atlantic Ocean. The tenacious Marconi pressed on after gale-force winds destroyed both initial towers. On December 12, 1901, during a lull in another hailstorm, the Italian inventor and his colleagues in St. John's received a Morse code message for the letter "S" that had been sent from Cornwall. Even with no independent witnesses, Marconi announced the achievement two days later to international acclaim; the New York Times pronounced this movement of messages through the air, without wires, "the most wonderful scientific development of recent times."[37]

An embittered Tesla argued Marconi's demonstration was predicated on his own oscillators, coils, and designs. The Italian scientist, in contrast, argued Tesla's equipment was ineffectual and unnecessary. Their arguments would rage for years and only accelerated after Marconi received the Nobel Prize in Physics in 1909.

Building on the laudatory press reports, Thomas Commerford Martin organized a celebratory banquet with three hundred scientific luminaries at the Waldorf's Astor Gallery, and he decorated the hall with a large map of the Atlantic Ocean, placed model transmission towers at each table, and draped an Italian flag around Marconi's portrait on the dais. Michael Pupin offered a qualified endorsement, saying "If I did not know him personally, I would not believe him, because the proof which Signor Marconi has furnished is not sufficiently strong from a purely scientific point of view." The Columbia University professor used the occasion to take a swipe at his fellow Serb, suggesting Tesla simply claimed he could transmit "wireless signals by the wobbling of the change of the earth" but never offered "an engineering specification of [his] apparatus."[38]

Despite the event being at his own residence, Tesla sent his regrets and a polite note: "I wish to join the members in heartily congratulating Mr. Marconi on his brilliant results. He is a splendid worker, full of rare and subtle energies. May he prove to be one of those whose powers increase and whose mind feelers reach out farther with advancing years for the good of the world and honor of his country."[39] Yet Tesla proved to be less tactful when the New York Times subsequently asked for comment. Suggesting his own innovative equipment and pioneering theories of wireless telegraphy predicated Marconi's accomplishment, the inventor moaned that the Italian had sent only a tiny message while bragging that he, Tesla, soon would reveal a more sophisticated and useful system.

The newspaper also requested a statement from Thomas Commerford Martin, editor of *Electrical World*, who praised Marconi and bluntly dismissed Tesla. "I am sorry," he said, "that Mr. Tesla, who has given the matter so much thought and experimentation, and to whose initiative so much of the work is due, should not also have been able to accomplish this wonderful feat."[40] Martin, however, went out of his way later to mend his relationship with Tesla, writing in a personal note: "I know of no change whatever in my sentiments towards you these many years from the beginning until now. I shall always be very proud of my modest association with your earlier work."[41] Relations improved a bit more in 1903 when Martin sent to Tesla a book he dedicated to the scientist, prompting Tesla to reply: "Many thanks for the book. It was a pleasure to read the dedication which tells me that your heart is true to Nikola."[42]

While Marconi amassed more and more acclaim, Tesla merely amassed more debt. He approached Morgan for a larger loan, but the financier refused and even delayed by two months making his final payment on the initial agreement. Tesla persisted with more letters regaling the wonders of his "World Telegraphy System" and its "tremendous money-making power."[43] His plan called for Wardenclyffe-like towers near "large centers of civilization" that "as fast as they receive the news, they pour them into the ground, through which they spread instantly."[44]

Tesla even described a handheld, inexpensive receiver that would allow someone to listen to anything virtually anywhere.[45] Well before the age of mass production, not to mention some seventy years before the introduction of cell phones, he proposed manufacturing millions of these devices.

The inventor defended his leadership on wireless transmissions, arguing that "all the essential elements of the [Marconi] arrangements . . . are broadly anticipated by my patents of 1896 and 1897."[46] He then dished out more idealism. "I have never attempted, Mr. Morgan,

Tesla at twenty-three.

37 UNION SQR, N.Y.

Tesla at twenty-five. *(Kenneth M. Swezey Papers, Archives Center, National Museum of American History, Smithsonian Institution)*

Tesla illuminating his wireless bulb.
(Smithsonian)

Lab at Colorado Springs. *(Smithsonian)*

Generators at Niagara Falls. *(Smithsonian)*

Court of Honor, Chicago World's Fair. *(Kenneth M. Swezey Papers, Archives Center, National Museum of American History, Smithsonian Institution)*

Tesla with King Peter II. *(Smithsonian)*

Wardenclyffe Tower. *(Smithsonian)*

Milutin, father of Nikola Tesla.

Tesla before his coil.

Tesla at thirty-four.

Tesla at forty.

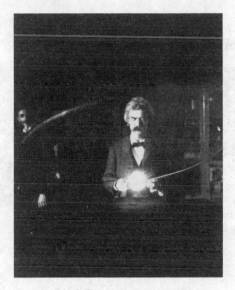

Mark Twain in Tesla's lab.

Tesla's lab in Colorado Springs.

Tela's alternating-current motor. *(Getty)*

Thomas Edison.

Tesla lecturing at Columbia College.

Robert Underwood Johnson.

J. P. Morgan.

Tesla's remote-controlled boat.

Katharine Johnson.

George Westinghouse.

to tell you even a hundredth of what can be readily accomplished by the use of certain principles I have discovered," he wrote. "If you will imagine that I have found the stone of the philosophers you will be not far from the truth. They will cause a revolution so great that almost all values and all human relations will be profoundly modified." The ever pragmatic Morgan, however, wanted to hear about financial returns rather than revolutionary visions.[47]

Tesla resorted again to flattery and pleading. "Now, Mr. Morgan, am I backed by the greatest financier of all times?" he asked. "And shall I lose great triumphs and an immense fortune because I need a sum of money?[48] He then begged: "Are you going to leave me in a hole?!! I have made a thousand powerful enemies on your account, because I have told them that I value one of your shoe-strings more than all of them."[49] He also added a little nationalism: "Is it not due to the honor of this country that it be identified with this achievement? Have I not contributed to its greatness and prestige and have my inventions not exercised a revolutionary effect upon its industries? These are not my claims, Mr. Morgan, only my credentials."[50]

Tesla later attempted more bravado. "I hope you will never for a moment confound my art with the incompetent efforts of my imitators. I could do better than any of them, if ninety-nine parts of me were paralyzed."[51] He also dismissed the profit potential of Marconi's efforts, suggesting "there was no money" in reporting yacht races or signaling incoming steamers. "This was no business for a man in your position and importance," he told Morgan.[52]

The persistent (and somewhat pathetic) scientist then issued threats. "What a dreadful thing," he wrote, "it would be to have the papers come with your name in red letters—A MORGAN DEAL DEFAULTS. It would be telegraphed all over the globe."[53] Tesla candidly admitted to Morgan: "Financially, I am in a dreadful fix."[54]

The inventor knew on some level that his pleas were ineffective. He admitted writing often "in moments of despondence with the pain [was] too hard to bear."[55] This once disciplined thinker, however, couldn't stop himself.

Hearing nothing, a frantic Tesla decided to go public, writing a six-thousand-word treatise for *Electrical World & Engineer* about the value of his ideas, hoping the financier would suddenly realize the potential and change his mind. The effort probably had the opposite effect, in part because the way he expressed his vision—that "the entire earth will be converted in a huge brain"—seemed outlandish and impractical. And the less-than-tactful Tesla ended his essay with a public jab at Morgan's judgment: "Perhaps it is better in this present world of ours that a revolutionary idea or invention instead of being helped and patted, be hampered and ill-treated in its adolescence . . . by selfish interest, pedantry, stupidity and ignorance; that it be attacked and stifled; that it pass through bitter trials and tribulations, through the heartless strife of commercial existence."[56]

When he still heard nothing, Tesla lashed out and criticized the "Great Man," declaring, "You are no Christian at all, you are a fanatic musoulman [Muslim]. Once you say no, come what may, it is no."[57]

Morgan finally replied, in his own handwriting, "I regret to say that I should not be willing to advance any further amounts of money as I have already told you. Of course I wish you every success in your undertakings."[58]

Tesla continued pleading, insensitive to Morgan's logical observation that Tesla's wireless dream was more costly and less practical than Marconi's achievement. "You fire me like an office boy," Tesla complained, "and roar so that you are heard six blocks away; not a cent. It is spread all over town, I am discredited, the laughing stock of my enemies."[59] Several months later, Morgan's assistant firmly responded: "Dear Sir, Referring to your letter of the 13th October, Mr. J. P. Morgan wishes me to inform you that it will be impossible for him to do anything more in the matter."[60]

During this period, Tesla tried to secure backing from numerous other financial and business magnates, including Henry Clay Frick (an industrialist who also lived in the Waldorf), H. O. Havemeyer (of American Sugar Refining Company), Jacob Schiff (a Wall Street banker), Oliver Payne (a colleague of J. D. Rockefeller), and John Sanford Barnes (president of the St. Paul and Pacific Railroad). They all rejected Tesla's proposal, often asking why Morgan wasn't investing more. Potential investors found Tesla's proposals to be "vague and nebulous." For instance, after Tesla's pitch to the Independent Lamp and Wire Company, a vice president noted the inventor "had no concrete ideas, much less practical data, which would indicate the direction of the research he wanted to conduct."[61]

Despite his complicated and strained history with General Electric, Tesla even made a presentation to its president, Charles Coffin, and later wrote: "If [the GE people] refuse they are simply snoozers." Nothing came of that meeting.[62] Thomas Fortune Ryan (an insurance and tobacco magnate) initially agreed to provide some financing, but it appears Morgan talked him out of it. Tesla then repetitioned George Westinghouse, who claimed to have become only the titular head of his company without access to investment funds.

Part of Tesla's problem was Morgan controlled 51 percent of his patent rights on any wireless transmission system. Since the financier would obtain a disproportionate share of any future profits, new investors anticipated little incentive.

Tesla also reached out to his family, including his previously supportive uncles as well as his two surviving sisters—Milka and Marica—each of whom had married a priest, which would suggest they had few resources to share. Since Tesla had sent funds to relatives when he was financially flush, he felt it appropriate to ask for some when stretched. The largest contribution, several thousand dollars, came from Nikola Trobojevic, Tesla's nephew who engineered a steering gear for which he obtained a royalty contract with General Motors. Yet when Trobojevic stopped the contributions after he lost his GM deal and his wife became ill, Tesla responded bitterly, "When I was in most dire straits you did not help. I cannot forget that you denigrated my integrity."[63]

A desperate Tesla continued to press J. P. Morgan: "You let me struggle on, weakened by shrewd enemies, disheartened by doubting friends, financially exhausted, trying to overcome obstacles which you yourself have piled up before me. Will you let a property of immense value be depreciated, let it be said that your own judgement was defective, simply because you once said no?"[64]

In a lengthy and lucid thirteen-page letter, Tesla described the history of his patents and plans. He admitted altering his initial agreement with Morgan to manufacture oscillators and cold lightbulbs. He claimed his switch toward wireless transmissions was necessary because the "audacious schemer" Marconi had pirated his radio patents and because he envisioned an opportunity to transmit far more than Morse-coded messages. Tesla acknowledged other financiers avoided him after he'd lost support from Morgan, who was understandably "skeptical of getting hundred-fold returns." He ended with a weak plea: "If you will help me to the end, you will soon see that my judgment is true."[65]

The financier himself replied "no" once again, declaring Tesla's trials were of his own making. "I am *not* willing to advance you any more money as I have frequently told you," Morgan wrote. "I have made and carried out with you in good faith a contract and having performed my part, it is not unreasonable that I expect you to carry out yours."[66]

Stoking an image of Tesla as a visionary and benevolent genius oppressed by entrenched interests, some historians have suggested Morgan blocked further investments for fear the inventor would devise systems that threatened the financier's stake in General Electric as well as in copper mines that supplied the wire for transmitting electricity. One such someone-said-to-someone-else suggestion came from Andrija Puharich, an inventor and physician who knew Tesla's friend and first biographer, John (Jack) O'Neill. According to Puharich, O'Neill "said that Bernard Baruch [a successful stock investor] told J.P. Morgan, 'Look, this guy is going crazy. What he is doing is, he wants to give free electrical power to everybody and we can't put meters on that. We are just going to go broke supporting this guy.' And suddenly, overnight, Tesla's support was cut off, the work was never finished."[67]

If this account is accurate, Tesla should not have been surprised by such a reaction from Morgan, having several years before revealed the economically disruptive nature of his inventions. Bragging his wireless system would provide cheap electrical energy throughout the world, he declared that "all monopolies" that depend on wires to distribute power—including utilities, electrical equipment manufacturers, and copper mine owners—"will come to a sudden end."[68]

Another explanation, as Morgan himself wrote, is that Tesla had not delivered and the financier was unwilling to waste more money. It had been two and a half years since Tesla promised to span the Atlantic, and it instead was Marconi who delivered on that promise. The financier mostly wanted a practical means to signal ocean liners and send Morse code to Europe, whereas Tesla kept envisioning a worldwide web of communications and power.

Morgan also had more important things brewing, from merging firms into International Mercantile Marine and International Harvester to defending himself against President Theodore Roosevelt's antitrust investigation of his Northern Securities Company. Perhaps more importantly, the stock market fall of 1903, referred to as the "Rich Man's Panic," disrupted financing for virtually all projects. Four years later, the Bankers' Panic of 1907 further prompted shares on the New York Stock Exchange to plummet, sparking another round of investment cutbacks.

Still, Tesla issued yet another missive, labeling Morgan as "the biggest Wall Street monster" who had broken their contract for what he labeled "the greatest invention of all times." He described the financier as "a big man, but your work is wrought in passing form, [while] mine is immortal." Egotistical and frantic, Tesla boasted: "I have more creations named after me than any man that has gone before not excepting Archimedes and Galileo—the giants of invention. Six thousand million dollars are invested in enterprises based on my discoveries in the United States today."[69]

Tesla childishly cranked up the power at Wardenclyffe, heaving streamers into the sky and prompting neighbors, according to a newspaper, to fret about "all sorts of lightning . . . from the tall tower. Until after midnight the air was filled with blinding streaks of electricity that seemed to shoot off into the darkness on some mysterious errand."[70] People in Connecticut, some twenty miles across the Long Island Sound, witnessed the artificial lightning and the cupola glow an eerie blue.

(In the 1930s, when Tesla was in his seventies and growing senile, he seems to have forgotten his frustrations with Morgan, bragging to his secretary that he "could get money from him just by asking for it." Supposedly, Tesla casually walked into the mighty banker's office and "Mr. Morgan asked if he could write out a check for me and called for the boy to bring his book. Morgan signed a blank check and asked me to fill in the amount I needed.")[71]

Finances weighed heavily on Tesla. "The obstacles in my way," he wrote to his assistant, "are a regular hydra. Just as soon as I chop off a head, two new ones grow."[72] In June 1904, the Colorado Springs power company won a $180 judgment that allowed them to tear down Tesla's lab and sell the lumber.[73] To make matters worse, after seventeen years Tesla's fundamental poly-phase patents expired in May 1905, reducing a source of royalties and payments from various manufacturers that had utilized his inventions.

Despite his financial woes, Tesla continued to live rather regally at the Waldorf Astoria. He shifted his regular dinners from Delmonico's to the hotel's grand Palm Room, where guests ate beneath an ornate three-storied dome of amber glass, from which hung a sculpted chandelier. The lavish dining room was filled with large palms and decorated in the Italian style and finished in terracotta and Pavonazzo marble. In order to be seen and keep up appearances, the well-dressed Tesla entered the restaurant each evening after walking through the glamorous, three-hundred-foot-long marble corridor that was nicknamed Peacock Alley. His table, at which he would dine regularly for eighteen years, stood near the wall and was set only for one.

The less-than-pragmatic inventor turned down a few lucrative contracts, including a request by Lloyd's of London to use his wireless system to report on an international yacht race. The inventor asserted he could not be bothered by minor projects.

"I confess my low commercial interests dominate me," he moaned to his friend Robert Johnson, whose home he continued to periodically visit for dinners and conversation. Tesla signed the letter "Nikola Busted."[74] He expressed humiliation in going from door to door "to solicit funds from some Jew or promoter." Frustrated and with few options, he cried, "I am tired of speaking to pusillanimous people who become scared when I ask them to invest $5,000 and get the diarrhea when I call for ten."[75]

(This was not Tesla's only anti-Semitic comment. His secretary remembered the inventor complaining about Fritz Lowenstein, a former assistant, making a profit from Tesla's radio inventions. According to Dorothy Skerritt, Tesla "drew close and whispered, as he always did when he had something important to say, 'Miss—Never trust a Jew!— Never trust a Jew!'"[76] A priest, moreover, recalled the inventor saying, "I am born idealist, not materialist; there is no Jewish blood in me."[77] In context, such comments were not uncommon among contemporary business and engineering leaders, and it doesn't appear Tesla discriminated against anyone based on religion. He later would rail against Hitler, although his criticism of fascists was limited to their treatment of Serbs.)

Claims on his limited money seemed limitless. James Warden demanded back taxes on the Long Island land, the phone company threatened to cut off service, and the Westinghouse Company wanted some thirty thousand dollars for its equipment at the laboratory. Since Tesla had failed to pay his bills and taxes, the Colorado Springs sheriff put his remaining electrical equipment there up for sale; an electrical transformer went "for about $10, or one-tenth of its market value."[78] To a friend Tesla declared, "I SWEAR if I ever get out of this hole, nobody will catch me without cash!"[79] Colleagues were becoming worried about his outlook. George Scherff, after one of the scientist's visits to Wardenclyffe, remarked: "I have scarcely ever seen you so out of sorts as last Sunday and I was frightened."[80]

The only thing that cheered Tesla up was a visit from Richmond Hobson, the war hero and naval engineer. In a letter to another acquaintance, Hobson revealed, "I went to see dear Tesla. He kissed me on the cheek, as once before, and when I left him at one o'clock last night, I felt he was prepared and ready for another year and for future years."[81]

About a year and a half later, in spring 1905, Hobson startled Tesla with news of his plan to marry Grizelda Houston Hull, from a well-connected Kentucky family. Hobson's letter revealed the close relationship between the engineer and the inventor. "Do you know, my dear Tesla," he wrote, "you are the very first person, outside of my family that I thought of . . . I wish to feel you present in standing close to me on this occasion so full of incoming in my life. Indeed, I could not feel the occasion complete without you. You occupy one of the deepest [places] in my heart."[82]

Tesla, a self-described celibate, tried to talk Hobson out of marrying. "Your career will be over once you marry," he argued—to no avail.[83] With some prodding, Tesla served as an usher in the wedding at the Hobson home in Tuxedo Park, New York. Yet it took several years, according to Grizelda Hobson, for Tesla to finally become "reconciled to his dear friend 'Hobson' getting married."[84]

(Hobson became a rear admiral in the U.S. Navy as well as a Democratic member of the U.S. House of Representatives from Alabama during the years 1907 to 1915. He and Tesla remained close for years, often going to dinner or the cinema. According to Grizelda, "These two dear friends, about once a month or sometimes oftener, would meet and go to a movie and then sit in a Park, and talk till after midnight! Richmond always came home with enthusiasm over some new invention of Tesla's and well I recall the night he told Richmond, 'I can shake the world out of its orbit, but I won't do it, Hobson!'" When Hobson was buried in Arlington Cemetery in March 1937, Tesla sent a "gorgeous flowering azalea" to Mrs. Hobson at her New York apartment.)[85]

By the fall of 1905, a stressed Tesla suffered "abused nerves (that) finally rebelled and I suffered a complete collapse."[86] He struggled for months with ominous dreams, particularly about the deaths of his brother and mother, finding those visions, he later wrote, "almost unbearable and every night my pillows were wet from tears."[87] The frustrated scientist lashed out bitterly at the "blind, fainthearted, doubting world."[88]

Nothing seemed to be going well for Tesla. In June 1906, his friend and colleague, Stanford White, was shot three times and killed at Madison

Square Garden by the jealous husband of a former chorus girl, Evelyn Nesbit, who was White's mistress. Perhaps fittingly, noting White's numerous dalliances, the millionaire husband, Henry Thaw, aimed his pearl-handled pistol between the eyes of White, whom he called "the Beast," while the entertainer at the Roof Top Garden sang "I Could Love a Thousand Girls." The scandal kept most colleagues and acquaintances from the funeral, but Tesla arose from his sickbed to attend.

Tesla had lost several key figures in his life, most notably Anthony Szigeti in 1890 and his mother in 1892, but the death of White, with whom he shared dinners at The Players and leisurely walks on Long Island, distressed the inventor during a rocky period and accelerated his sense of isolation. "I'm ever in so much trouble," he confessed in erratic penmanship to his friend Katharine Johnson.[89] He blocked all sunlight from entering his hotel room. He refused to eat. Electrotherapy to his brain became his only treatment. "I have passed [150,000 volt]) through my head and did not lose consciousness," he wrote, "but I invariably fell into a lethargic sleep sometime after."[90] Sadly, his ongoing trials prompted Tesla to abandon some of his ambitions and to temper his drive. "The practical lesson of all this," he wrote, is "to beware of concentration and be content with mediocre achievement."[91]

12.

TOO MUCH OF A POET
AND VISIONARY

New York City

Tesla slowly recovered his strength but not his Wardenclyffe project. Debts mounted, the Port Jefferson Bank demanded loan repayments, and workers trickled away or sued him for back wages. On a foggy morning, the Westinghouse Company, bolstered by a court order, sent horse-drawn wagons to collect its equipment. Tesla described that day as the saddest of his life. A newspaper referred to the Long Island venture as "Tesla's million-dollar folly."

In April 1912, when John Jacob Astor died aboard the *Titanic*, Tesla lost another key contact. Although Astor had stopped supporting Tesla's research years before, he'd allowed the scientist to live rent-free in the Waldorf Astoria. By 1915, Tesla was forced to sign the Long Island property's deed over to the Waldorf Astoria Company, whose new proprietor had demanded three years of rent. The hoteliers failed to convert the Wardenclyffe property into a tourist attraction, and the War Department rejected a plan to use the tower for reconnaissance against enemy submarines. Some unverified news reports suggested German spies deployed the structure to communicate with their ships off the U.S. coast. By July 1917, a salvage company dynamited the tower—multiple times because of its sturdy construction—and sold the scrap for only $1,750.

Bringing down the tower also brought down Tesla's dream of delivering wireless communications and power to the world. Bitter and frustrated, he blamed "small-minded and jealous individuals" who "are to me nothing more than microbes of a nasty disease." Despite overreaching on the science and under-reaching on funding, he was right to declare, "The world was not prepared for it. It was too far ahead of time, but (it) will prevail in the end and make it a triumphal success."[1]

The financial stress led to a change in Tesla's tone, expressing a more frequent willingness to belittle and boast. He started calling his competitors "feeble men," including Professor Ferraris, even though, as noted earlier, he had admitted "Tesla developed [an AC poly-phase system] much further than [I] did."[2] Tesla increasingly discounted Edison and Marconi as pioneers, particularly compared to himself: "I had to cut the path myself, and my hands are still sore." His references to his own work became more and more vain, as when he insisted his system to send power wirelessly through this planet "is so perfect now that it admits of but little improvement. . . . Would you mind telling a reason why this advance should not stand worthily beside the great discoveries of Copernicus?"[3]

No doubt Tesla had reasons to be bitter. He had invented the induction motor, the poly-phase system for transmitting alternating current, remote control, robots, wireless communications, and the "individualization" of messages. While a slew of businessmen using his insights had made millions, he lamented not having received his due in credit or compensation. In a letter to the *New York World*, he complained: "Had the Edison companies not finally adopted my invention they would have been wiped out of existence, and yet not the slightest acknowledgment of my labors has ever been made by any of them, a remarkable instance of the proverbial unfairness and ingratitude of corporations."[4] Tesla also lashed out at the Westinghouse Company for allegedly pirating his wireless patents for its radio machinery. "You have robbed me of the credit that is due and injured me seriously in business," he wrote to the firm's vice president. "Instead of showing a willingness to adjust the matter in an equitable way, you say you want to fight."[5] The embittered inventor was becoming increasingly isolated.

Adding to that loneliness was the death of George Westinghouse in March 1914 in New York City at the age of sixty-seven. (Tesla then was almost fifty-eight.) As a Civil War veteran, Westinghouse was buried at Arlington National Cemetery. Despite Tesla's periodic complaints against the company, George Westinghouse had perhaps been his most stalwart supporter. The firm had for many years been easing out its founder and that process culminated with the financial panic of 1907 when financiers forced Westinghouse's resignation. Before his ouster, George had remained an active inventor, developing steam turbines, shock absorbers for automobiles, and heat pumps that provided both heating and cooling.

Despite being financially strapped, more and more alone, and mentally unstable, Tesla, even in his late fifties and early sixties, kept his looks and agile build. His weight did not vary by even a single pound for almost thirty-five years, and his suits continued to fit him "like a glove." When the inventor was fifty-nine, one reporter observed: "Where he keeps his years, no one knows. They are not in his face, for his face looks like forty. They are not in his hair, for his hair is black. If they are anywhere, they are in his eyes, for his eyes are sad."[6] To demonstrate his strength and agility, Tesla recalled, probably with a good bit of exaggeration, walking back to his New York hotel on a cold, slippery night. When he lost his balance and his legs went out from under him, there was, he said, "a flash in my brain, the nerves responded, the muscles contracted, I swung 180 degrees and landed on my hands." Brushing himself off, he resumed his march. A man walking behind him marveled at Tesla's feat. "'How old are you?' he asked critically. When told fifty-nine, according to Tesla, the man observed, 'I have seen a cat do this, but never a man.'"[7]

Without a major donor, Tesla finally was forced to take on varied projects, such as working with Budd Manufacturing of Philadelphia on a petrol-powered turbine; in Detroit, he helped Ford, General Motors, and other automakers design a flying car. He spent nine months in Chicago trying to develop headlights for railroad engines. In contrast to his early claims of perfecting devices in his mind, he now admitted,

"As we advance we recognize the crudity of our first conceptions. New problems presented themselves which had to be solved and they were so hard that they consumed about all the energy I could command."[8] Such projects allowed him to hire an assistant or two, but the funds were never enough to restart another Wardenclyffe.

More problematic, scientists began to realize Tesla was wrong about the earth's ability to transmit messages and power. Nikola had assumed electrical power would flow through the planet and be available to receivers atop it. Comparing the earth to a balloon (and sometimes to a metal ball), he believed the globe's interior would act like water, or any incompressible fluid, where waves flow back and forth with little loss of power. Yet the earth's interior behaves more like a giant ocean, with energy waves dissipating and eventually disappearing. Put another way, although Tesla never acknowledged it, the earth proved to be an inefficient conductor of electricity or messages.

Another fundamental error was Tesla's belief that power and telegraphic engineering needed complete circuits. Focusing on the electrostatic thrusts from his oscillator, he largely dismissed the electromagnetic theory advanced by James Clerk Maxwell in 1865, proved by Heinrich Hertz in the late 1880s, and promoted in the 1890s by so-called Maxwellians—including Oliver Lodge, John Perry, and George Francis FitzGerald. We now know that these waves of electromagnetic field—including visible light, gamma radiation, and radio waves—radiate through space and don't require electricity's return circuit. Being wrong periodically, of course, is a natural part of inventing. Thomas Edison, for instance, spent substantial sums and time trying to improve the iron-ore milling process, but his technological developments proved to be unprofitable. The Wizard of Menlo Park, however, carried on, commenting on his financial losses: "It's all gone, but we had a hell of a good time spending it."[9]

Tesla tried to carry on as well. He reluctantly acknowledged the criticism from those he called "my enemies" that he was too much of "a poet and visionary" rather than the producer of "something commercial,"[10] yet in his heart, he most enjoyed imagining the ideal. Age was tempering the precision of his mind, yet he also had tilted his balance toward tomorrow and visions. His thoughts progressively became viewed as prophetic or delusional, depending on the viewer's perspective. Some criticized him as a scientific huckster, but there's no doubt he continued to think deeply and broadly about a range of challenges and opportunities.

Foreseeing the Internet and even the smart watch, for instance, he predicted in 1908 "it will be possible for a business man in New York to dictate instructions, and have them instantly appear in type at his office in London or elsewhere. . . . An inexpensive instrument, not bigger than a watch, will enable its bearer to hear anywhere, on sea or land, music or song, the speech of a political leader, the address of an eminent man of science, or the sermon of an eloquent clergyman, delivered in some other place, however distant. In the same manner, any picture, character, drawing, or print can be transferred from one to another place."[11]

Tesla got many things right; however, he also made predictions that proved to be embarrassing. In 1908, five years after the Wright brothers' demonstration in North Carolina but nineteen years before Charles Lindbergh's transatlantic flight, he proclaimed that the "aeroplane is altogether too heavy to soar." The better alternative, he reasoned, was the dirigible balloon being developed by Count Zeppelin that was "safe and reliable, [able] to carry a dozen men and provisions, and with a speed far in excess of those obtained with aeroplanes."[12] Yet even with this miscalculation, Tesla offered the insight that "the propeller is doomed" at high velocities and would be replaced by "a reactive jet."[13]

While Tesla prophesized, other inventors produced. Lee de Forest, for instance, boosted the speed of Morse-code transmissions to six hundred words per minute, and he signed a contract with Bell Telephone to build a "radio wireless" system between New York and Philadelphia. (De Forest as a young man had asked Tesla for a job, but the inventor turned him down, predicting he needed his independence in order to achieve great things. De Forest went on to develop the three-element "Audion" vacuum tube that enabled radio broadcasting and became the basis for electronics, but he maintained throughout his life that Tesla had been his greatest inspiration.)

Perhaps Tesla's most ingenious product during this challenging period was a bladeless turbine that was to create a more efficient electrical system and more powerful airplane engines. Rather than rely on the complex rotating blades or buckets used in conventional turbines, which resembled windmills within chambers, Tesla designed a machine that would press fluid, steam, or air through a series of carefully racked disks, and as the fluids or gases spiraled down, they dragged and rotated a shaft with them. In reverse, the fluid or gas would spiral out from the center and act as a pump.

Tesla, returning to a theme underlying many of his inventions, claimed his goal for the turbine was "simplicity itself." The machine, he declared, would have "no exciter, no commutator, brush or sliding contact whatever, no centrifugal regulator, voltage controller or any such complicated and hazardous devices." By "its adaptability to high temperatures far beyond those practicable in bucket turbines," Tesla's device promised significantly higher efficiencies. The innovative turbine, moreover, would "adjust itself instantaneously to pressure changes... without the slightest observable change in the intensity of the light."[14]

From 1911 to 1918, he devised several such turbines—ranging in size from one hundred to five thousand horsepower—at the Waterside Power Station of the New York Edison Company. Their key advantage proved to be the ability to produce far more horsepower per pound of weight, making them lighter and smaller, yet more powerful, than bladed turbines. "I have an engine that will give ten horse power to a pound of weight," the inventor explained. "This is twenty-five times as powerful as the lightest weight engines in use today."[15] The engine, moreover, could be small and portable, what he called "a powerhouse in a hat."[16]

Tesla again demonstrated his unique perspective on nature. While most engineers wanted to limit the "skin friction" that put a drag on boats and vehicles, the inventor sought to make use of this force, what he called "viscous shear." He understood that gases and liquids can be sticky substances. If you pour water on a smooth surface, for instance, most will roll off but some will stick and cause resistance. By "transforming that old hindrance into a new help," one journalist observed, "Tesla's turbine balances a wheel in a current of steam or gas in such a manner that it is caused to rotate, not by the push but by the pull of the steam."[17]

For at least a moment, Tesla could again bend nature to his will. He creatively envisioned harnessing negative forces to his advantage. Unfortunately, he couldn't turn this ideal into something practical.

This scientist who'd been known as a master showman failed to devise a convincing demonstration. Instead, his tug of war between two linked turbines produced loud straining but no motion. Unimpressed audiences declared the test to be unsuccessful.

On the business front, the independent inventor could not locate a manufacturer willing to accept the risks associated with producing a novel machine. The major producers of turbines, Westinghouse Manufacturing and General Electric Company, were not interested in a design that might compete with their bladed turbines. "I have been striving ... to have a strong and well equipped manufacturing establishment putting on the market improvements as I may make them," Tesla wrote in May 1918. When he targeted a small manufacturer, the inventor noted "they have extraordinarily efficient personnel—mostly young and aggressive men and the factory is large, quite modern, and up to date."[18] Yet he soon realized the industrialist had interests other than his turbine, and he disappointingly concluded they were "awfully busy people, hustlers who are away most of the time."[19]

On the technical front, the shaft's high speeds put significant stress on the thin disks, often causing them to warp. Steel alloys strong enough to withstand the strain had not yet been developed. According to one engineer several years later, "Tesla was about twenty-five to thirty years ahead of his time. Metallurgy was not what it is today. Magnetic bearings are a whole new science. He didn't have the right materials."[20]

Tesla, once again, proved to be his own worst enemy. When he finally discovered interest from the Allis-Chalmers Manufacturing Company of Milwaukee, he demanded to work only with the firm's president, ignoring the engineers who had to build and test the device.

When those piqued workers pushed for several modifications, Tesla simply walked away and haughtily declared, "They would not build the turbines as I wished."

The bladeless turbine would have been considered a significant advancement if modern metallurgy had been available in the early twentieth century. Still, the device, with its process reversed, operated efficiently as a pump, and Tesla sold a license to the Alabama Consolidated Coal & Iron Company. With rediscovered optimism and humor, Tesla ordered from a friend at the Westinghouse Company a million induction motors to drive his new turbines, though he jokingly admitted, "But as I have learned to go slow, I shall take only one at first."[21]

Tesla also suggested the basic concepts for what became known as radar. In order to detect enemy aircraft and ships, he proposed in August 1917 to "shoot out a concentrated ray comprising a stream of minute electric charges vibrating electrically at tremendous frequency, say millions of cycles per second, and then intercept this ray, after it has been reflected by a submarine hull for example, and cause this intercepted ray to illuminate a fluorescent screen on the same or another ship, then our problem of locating the hidden submarine will have been solved."[22] Seventeen years later, when Tesla was still alive, Dr. Emile Girardeau and his French team built and installed radar on ships and land stations, using, he said, "precisely apparatuses conceived according to the principles stated by Tesla."[23] Girardeau added that Tesla may have been "prophesying or dreaming, since he had at his disposal no means of carrying them out, but one must add that if he was dreaming, at least he was dreaming correctly."[24]

Another prophetic idea was for a small but powerful winged aircraft that could rise like a helicopter straight into the air and then proceed straight like an airplane. With a wingspan of eight feet, the five-hundred-pound vehicle would be powered by a turbine engine and be capable of carrying two people. The pilot could tilt the propeller "through manipulation of the elevator devices,"[25] and his seat would swivel to remain upright as the wings moved into horizontal positions. The aircraft would land by reversing these processes. Tesla's vertical-takeoff-and-landing (VTOL) proposal, for which he obtained a patent in 1928 (which would be his last filing, at the age of seventy-two), was the first known design for a VTOL. Although no one was willing to finance its production, Tesla's plan became several decades later the basis for aircraft manufactured by aerospace giants for the United States Navy and Marines.

Tesla missed out on practical opportunities to develop his robots or telautomatons, largely because of his stubbornness and egocentricities. The wealthy Hammond family, which had given ten thousand dollars to Tesla in 1898 for the advancement of selective tuning devices, wanted to form the Tesla-Hammond Wireless Development Company, as Jack Hammond explained, in order to "perfect an automatic selective system, to perfect the [submersible] torpedo, and eventually to carry out your magnificent projects that will wirelessly electrify the world."[26] Hammond maintained the scientist had filed a "prophetic genius patent" in 1903 that allowed for combinations of frequencies to send specific directions to machines or lights. The signal-sending devices, analogous to modern television scramblers, ensured privacy as well as allowed for a virtually unlimited number of signals.

Jack Hammond should have been a perfect match for Tesla. He began his own inventing at a New Jersey prep school at the age of sixteen, when he designed a reverse switch that turned off his light when the headmaster opened his own door to check on curfew violators. While an engineering student at Yale University, he became interested in remote control and found Tesla and Bell to be "my scientific godfathers." Hammond would accumulate more than eight hundred patents from the United States and other countries, became known as the "father of radio control," and laid the foundation for modern missile guidance systems.[27]

Hammond, moreover, admired Tesla as a true inventor and claimed his approach could neither be duplicated by regimented research nor by the simple expenditure of millions of dollars. "It takes a profound personal dedication for a man to achieve high inspiration," Hammond concluded, "and this Nikola Tesla had."[28]

Tesla, however, feared Hammond was stealing his telautomatics ideas. After a newspaper reported that Hammond was demonstrating remote-controlled torpedoes to the military, Tesla sent a bad-tempered note: "I think that you are playing a wireless possum. Notwithstanding your assurances, I will watch your progress and bring a friendly suit for infringement as soon as I ascertain that you are in funds."[29]

Hammond subsequently praised Tesla for having, a decade earlier, questioned the power of Hertzian waves and demonstrating "that waves propagated at a transmitting station travelled along the ground as a conductor." Yet Tesla fretted that Hammond gave too much credit to his archrival Marconi for developing "a complete and practical system of space telegraphy."[30] Hammond's wealthy father, meanwhile, didn't trust Tesla, sensing he "tends to spend gold as if it were copper."[31]

Still, the Hammonds offered Tesla substantial additional invest-ments, but they pleaded for the telautomaton or robot to be evaluated independently. Jack's brother wrote to the inventor: "As you know, we have advanced a great many thousands of dollars in the development of this [device] and have expected each week the past year to be in a position to have tested it." Tesla essentially ignored the appeals, still far more interested in wireless power transmission, despite the technical problems associated with it. The Hammonds persisted, highlighting the "splendid opportunity of having it thoroughly and honestly tested by peo-ple who would be the greatest benefit to us should these tests be success-ful."[32] Tesla, wanting more funds without any progress reports, haughtily responded, "The sacrifices which I have been compelled to make and the losses which I have suffered are such that if I were dealing with a man less attractive to me than yourself, I would disdain to answer."[33]

So once again, despite his financial struggles, Tesla walked away from substantial backing, as well as from the chance to actually develop his wireless system, in part because he couldn't work well with oth-ers and largely because he remained tunnel-focused on rebuilding his doomed tower. The Hammonds, as good businessmen, got their revenge by submitting their own wireless patents just after Tesla's ran out. They made millions and Tesla once more failed to profit from his invention.

The inventor's arrogance was becoming a noticed trait. A respected journalist who interviewed Tesla for *Scientific American* said the inventor had turned into a "temperamental genius."[34] A business colleague complained, "I have had contact with people who found him so difficult to deal with and so inconsistent and unreliable in his business relations and engagement."[35]

Tesla continued to respond to the growing criticism by arguing he simply was ahead of his time. "Perhaps I was a little premature [with the wireless delivery of electricity]," he told a group of reporters. "We can get along without it as long as my poly-phase system continues to meet our needs. Just as soon as the need arises, however, I have the system ready to be used with complete success."[36] Despite facing fundamental problems, Tesla persistently maintained that the wireless transmission of power was "the greatest invention of all time."[37]

Tesla's confidence in his abilities, in fact, rarely flagged. With each new idea, he wrote something like: It "cannot help but prove a colossal success." Finding financing, however, remained both a challenge and a fantasy. "The only trouble is where and when to get the cash," he often said, "but it cannot last very long before my money will come in a torrent."[38]

With little revenue and dwindling energy, Tesla in 1928 closed his office, retired his longtime secretaries, Dorothy Skerritt and Muriel Arbus, and sent thirty trunks of correspondence, papers, and models to the basement of the Hotel Pennsylvania (where they remained until moved again in 1934 to the Manhattan Storage Warehouse located at Seventh Avenue and 52nd Street).

The sorting of Tesla's financial woes fell to George Scherff, the manager whose expanding responsibilities included accountant, advisor, and fundraiser. Most of their daily exchanges involved business information, with Scherff forwarding notes similar to: "If you will kindly send me the information regarding royalty earned by the Nikola Tesla Company . . . I can prepare the tax report and will then call on you to have the same executed."[39]

Yet Scherff had also loaned money to Tesla (some biographers suggest forty thousand dollars, but that's equal to almost one million dollars today, which seems impossibly high for an accountant/office manager), and his repayment requests became increasingly anxious. Although Tesla's secretary once said the inventor "seemed to have Mr. Scherff hypnotized,"[40] Scherff pleaded: "I have bought a house in the country and am somewhat in financial difficulty. If you could let me have a small payment on account, it would aid me very much at this time."[41] In another note, Scherff said his creditors "are hounding me hard, and anything that you can do for me will be very much appreciated."[42] When Scherff complained that the scientist had forwarded a check for Mrs. Schwartz, a minor investor, rather than for him, Tesla scoffed, "I am sorry to note you are losing your equanimity and poise. Mrs. Schwartz is weak, and you are fully able to fight your own battles. You must pull yourself together and banish the evil spirits."[43]

Most of the time, however, Tesla blithely responded to Scherff's pleas with his trademark optimism: "I have great prospects which I expect to materialize during the next two or three weeks. As soon as they do I shall communicate with you and you may rest assured that I shall do all in my power to meet the situation."[44]

Scherff needed more than Tesla's hopefulness. "It is impossible for me to exist without a steady income," he finally wrote. "In view of your statement made week before last, that you might perhaps suspend operations here, I am very sorry to inform you that I am compelled to seek other employment."[45] Still, the two remained close, with Tesla writing letters of recommendation and Scherff continuing to complete the scientist's financial reports while working full-time for the Union Sulphur Company.

In fact, Scherff was one of the few colleagues to maintain contact as Tesla aged and his mental health declined. The inventor's penmanship became more and more sloppy and cross-outs began filling his letters; nonetheless, the accountant tried to cheer up his former boss with regular messages: "I hope sincerely that your trials will soon have an end and that in the coming year you will see your labors rewarded."[46]

Tesla increasingly failed to keep up with—or accept—advances made by modern scientists, including Albert Einstein, who was more than two decades younger than Tesla. Tesla wrongly argued cosmic rays and radio waves could move faster than light. He claimed to have "split atoms but no energy was released," and he declared "the idea of atomic energy is illusionary."[47] Rejecting the premise of Einstein's $E = mc^2$ formula, Tesla asserted, "The idea that mass is converted into energy is rank nonsense."[48] He also repeatedly scoffed at Einstein's theory of relativity, declaring, "I am absolutely ignoring every one of the tenets of the relativity theory, which to me is just a mass of error.... The relativity theory wraps all these errors and fallacies and clothes them in a magnificent mathematical garb, which fascinates, dazzles, and blinds people to the underlying errors. The relativity theory is like a beggar clothed in purple whom the ignorant take for a king."[49] (Einstein, despite such critiques, used Tesla's seventy-fifth birthday celebration to pay tribute to the inventor as "a successful pioneer in the field of high frequency currents.")[50]

Tesla maintained a nineteenth-century view of physics, rejecting the role of electrons and believing electrical energy moved through an "ether" that permeates space. While Einstein declared "the ether cannot be detected" and was not needed to explain how light travels through space, Tesla never abandoned his belief in the medium and maintained space cannot be curved because "something cannot act upon nothing."[51]

It is odd that this man who was born and lived his life partially in the future refused to consider, let alone embrace, new visions of nature. Although he kept foreseeing modern developments, the aging Tesla became less and less willing to view the world from new angles.

Tesla's reality became his lack of money, particularly when World War I stopped his royalty payments from European manufacturers. New York City in 1916 sued him for $935 in back taxes. After admitting under oath that he was penniless and living on credit, Tesla had the following exchange with Justice Finch:

"How do you live?" asked the court.

"Mostly on credit," replied Tesla. "I have a bill at the Waldorf that I have not paid in several years."

"Are there other judgments against you?"

"Scores of them."

"Does anybody owe you any money?"

"No, sir."

"Have you any jewelry?"

"No, sir; jewelry I abhor."

"Any automobiles?"

"No, sir."

"Any horses?"

"No, sir."[52]

Finding Tesla "possessed no real estate or stocks and that his belongings, all told, were negligible," the judge ordered a receiver to manage the scientist's affairs. Perhaps more embarrassing to Tesla was the fact that the *New York Times* and several other journals reported the news. The proud inventor tried to keep up appearances, but according to one account, "this event would mark the turning point in his life. Now he began the slow but steady turning away from society."[53]

The New York City lawsuit prompted others. Although resulting in few actual financial losses for Tesla, they diverted his attention and accelerated his anxiety. If he could have found humor in his stressful state, he might have laughed at a Mrs. Tierstein who was confined to an asylum and sued him for "throwing electricity at her."[54]

During this period, Robert Johnson's finances also declined—largely because *The Century Magazine* faced increased competition—to the point he asked Tesla for money, claiming, "I am holding on to my house by the skin of my teeth and desperately in need of cash."[55] Tesla tried to cheer up his friend, writing: "Do not worry about finances. Remember while you sleep I work and am solving your problems."[56] A few days later, the inventor expanded: "Please take my words seriously and do not worry and write your splendid poetry in perfect serenity. I will do away with all difficulties which confront you."[57] Yet Tesla couldn't hide his own difficulties, to which Johnson, when he finally realized the depth of his friend's plight, responded, "I did not know how tenuous things are with you. Do not think me unfeeling, and I shall let my own troubles go hang, rather than write you again on the subject. We can sympathize with each other at least."[58] It's telling that Tesla's best friend was a poet who also had few money-management skills.

To obtain a little cash, Tesla prepared a detailed prospectus, printed on vellum paper and adorned with a red-wax seal and his initials, advertising his "professional services in the general capacity of consulting electrician and engineer." He also took on odd jobs, including a controversial effort to improve student performance by releasing high-frequency charges into classrooms. Working with the superintendent of New York City's public schools, he deployed his coil in a pilot project to excite a group of "fifty mentally defective schoolchildren," hoping the pervasive energy would increase their aptitude test scores and introduce "a new era of education."[59] He tried a similar approach at Broadway theaters, creating highly charged dressing rooms in order to stimulate performers before they went on stage.

13.

SO FAR AHEAD
OF HIS TIME
New York City

Despite his age and financial problems, Tesla attracted more and more publicity. The *New York Times* in early November 1915 mistakenly suggested Tesla and Edison would share the Nobel Prize in Physics. Although he had heard nothing directly from the Nobel Committee, Tesla declared "the honor has been conferred upon me in acknowledgment of a discovery announced a short time ago which concerns the transmission of electrical energy without wires. This discovery means that electrical effects of unlimited intensity and power can be produced, so that not only can energy be transmitted for all practical purposes to any terrestrial distance, but even effects of cosmic magnitude may be created."[1] Edison, in contrast, wisely avoided making any statement.

When it became clear the newspaper report was false, Tesla claimed he would have refused the award since he wouldn't share it with Edison and couldn't accept a prize already given to Marconi. Hearing such, the Nobel Committee went out of its way to declare: "Any rumor that a person has not been given a Nobel Prize because he has made known his intention to refuse the reward is ridiculous."[2] A few weeks later, the panel formally announced the physics award for 1915 would go to Sir William Henry Bragg and his son William Lawrence Bragg for their use of X-rays to analyze the structure of crystals. Despite their genius and accomplishments, neither Edison nor Tesla would ever be recognized by the Nobel Committee.

Tesla was not gracious upon hearing of the panel's decision. To a *New York Times* reporter, he mockingly complained, "A man [Sir Bragg] puts in a kind of gap [into my coil]—and he gets a Nobel Prize for doing it. . . . I cannot stop it."[3] In a letter to a friend, he boasted, "In a thousand years, there will be many recipients of the Nobel Prize, but I have not less than four dozen of my creations identified with my name in the technical literature. These are honors real and permanent, which are bestowed, not by a few who are apt to err, but by the whole world which seldom makes a mistake and for any of those I would give all the Nobel Prizes which will be distributed during the next thousand years."[4]

The following year Tesla, then aged sixty-one, received an actual award, the Edison Medal from the American Institute of Electrical Engineering. With some testiness, he initially rejected the honor because it was named after his sometimes nemesis and because the AIEE had for so long ignored his achievements. "It is nearly thirty years since I announced my rotating magnetic field and alternating-current system before the Institute," he wrote to Bernard A. Behrend, the distinguished Westinghouse engineer who was pushing Tesla to accept the AIEE's highest accolade. "I do not need its honors and someone else may find it useful."[5] The persistent Behrend, however, eventually convinced Tesla to attend the white-tie ceremony at the Engineers' Club and the subsequent tribute across the alley at the United Engineering Societies. Behrend also needed to convince a few of the AIEE engineers to attend since, according to one member, "The stories of Tesla's sexual episodes were at one time the talk of the Institute, and we didn't know how to deal with it if the matter should somehow become publicized."[6] (No other members backed up that claim, and sexual exploits seem unlikely for a germophobe unwilling to touch anyone.)

A large crowd gathered on the night of May 18, 1917, including appreciative scientists and his dear friends Robert and Katharine Johnson, Marguerite Merrington (his frequent dining companion), and Edward

Dean Adams (his early financial backer). Tesla appeared at the initial gala to be alert, engaged, straight backed, and well-dressed. Yet somewhere on the way to the presentation event across the alley, the guest of honor disappeared. Waiters checked restrooms. Assistants investigated nearby buildings.

Eventually, on a hunch, Behrend decided to look in nearby Bryant Park, in which he remembered Tesla frequently took walks. There, in the shadows, stood the dapper and famous scientist—covered in pigeons. Birds fed from his outstretched arms. They sat on his head. They swarmed across his black evening pumps. Behrend quietly and carefully removed the creatures, and the two men walked slowly back to the United Engineering Societies building on 39th Street.

As if nothing had happened, Behrend offered his formal, though delayed, testimonial. "So far reaching is his work that it has become the warp and woof of industry," he said of Tesla. "His name marks an epoch in the advance of electrical science. From that work has sprung a revolution." Paraphrasing Alexander Pope's line about Sir Isaac Newton, Behrend concluded: "Nature and nature's laws lay hid by night. God said, 'Let Tesla be and all was light.'"[7]

When Tesla rose, he offered gracious comments about other inventors, particularly Edison, whom he called "this wonderful man, who had had no theoretical training at all, no advantages, who did all himself, getting great results by virtue of his industry and application."[8] He humbly suggested other speakers had "greatly magnified my modest achievements," and expressed his determination to "continue developing my plans and undertake new endeavors." Reflecting on his own life, he rambled a good bit and declared: "I have managed to maintain an undisturbed peace of mind, to make myself proof against adversity. . . . I have fame and untold wealth, more than this, and yet—how many articles have been written in which I was declared to be an impractical unsuccessful man, and how many poor, struggling writers have called me a visionary? Such is the folly and shortsightedness of the world!"[9]

How much did Tesla believe of what he said? One minute he was communing with pigeons as his own award ceremony took place and the next he was telling those gathered that he had "fame and untold wealth."

Tesla then spoke at length about his inventing process. He initially contrasted this "new method" to Edison's consistent "improving and reconstructing" until "your force of concentration diminishes and you lose sight of the great underlying principle." Instead, Tesla did "not rush into constructive work. When I get an idea, I start right away to build it up in my mind. I change the structure. I make improvements. I experiment. I run the device in my mind. . . . In this way, you see, I can rapidly develop and perfect an invention, without touching anything. When I have gone so far that I have put into the device every possible improvement I can think of, that I can see no fault anywhere, I then construct this final product of my brain."[10]

Tesla also used the occasion to reflect on mortality. "I come from a very wiry and long-lived race," he said. "Some of my ancestors have been centenarians, and one of them lived 129 years. I am determined to keep up the record and please myself with prospects of great promise. Then again, nature has given me a vivid imagination."[11] (Vivid indeed, since there's no evidence a relative of his lived 129 years.)

Tesla promised to continue battling the "darker side of life, the trials and tribulations of existence," by trying to maintain "an undisturbed peace of mind . . . and to achieve contentment and happiness."[12] And he kept inventing. In 1918, he licensed the world's first air-friction speedometer to the Waltham Watch Company, which installed those devices in luxury cars such as the Cadillac, Rolls-Royce, and Pierce-Arrow; acknowledging the aging inventor's continued popular appeal, Waltham mentioned Tesla's name in its advertisements.[13] The scientist also devised other meters to measure frequencies and flows, and he developed an innovative process for refining metals that he sold to American Smelting and Refining Company, a copper-mining giant. Inventing inspired Tesla, even during tough financial times. In the first segment of his serialized autobiography, published in 1919 in the *Electrical Experimenter*, he boasted, "I have already had more than my full measure of this exquisite enjoyment, so much that for many years, my life was little short of continuous rapture."[14] While most workers suffered an idle boredom or faced rigid and repetitive tasks, Tesla "thrived on my thoughts." The only downside of being a prolific inventor, he said, was that "so many ideas go chasing through his brain that he can only seize a few of them as they fly, and of these he can only find the time and strength to bring a few to perfection. And it happens many times that another inventor who has conceived the same ideas anticipates him in carrying out one of them. Ah, I tell you, that makes a fellow's heart ache."[15]

Some friends were becoming impatient with Tesla's aching heart. Hugo Gernsback, editor of *Electrical Experimenter*, had helped Tesla serialize his memoirs and provided needed attention and funds during this tough period, yet when Gernsback suggested featuring Tesla in a new futuristic magazine, the increasingly irritable scientist complained the monetary reward was too small: "I appreciated your unusual intelligence and enterprise, but the trouble with you seems to be that you are thinking only of H. Gernsback first of all, once more, and then again."[16] When Gernsback, who also published science fiction, asked Tesla to write an article on earth conduction, the inventor huffed: "Your statement that an article from me would be worth to your paper $100 might be flattering if my friend . . . had not offered me $2500 for two hundred words. On this occasion, I painfully remember that my contributions to the *Electrical Experimenter* cost me a few thousand dollars."[17]

Tesla's touchiness resulted in part from the consistent need to protect his inventions through the legal system. Believing his were the first patents and articles explaining the concepts for radio communications, he filed patent infringement lawsuits against Marconi in August 1915. Telefunken, the German telephone company, also sued Marconi, and the following year, the Marconi Wireless Telegraph Company of America responded by charging the U.S. government for its alleged unauthorized use of wireless during World War I. That charge prompted Franklin Roosevelt, then acting secretary of the Navy, to suggest the government's case against Marconi would be strengthened by Tesla's old correspondence with the Light House Board that demonstrated Tesla's "ability to supply wireless telegraph apparatus."[18]

The first case was adjudicated before France's highest court. Tesla testified that Marconi's patent of June 2, 1896, was "but a mass of imperfection and error.... If anything, it has been the means of misleading many experts and retarding progress in the right direction."[19] Judge M. Bonjean agreed, declaring Tesla's invention had come before Marconi's.

The government's lawsuit wound through the U.S. courts for years, with a Brooklyn district judge siding with Marconi and then the Court of Claims in 1935 overturning that ruling, invalidating the Italian's fundamental patent, and claiming it had been preceded by Tesla. Almost another decade would pass before the U.S. Supreme Court weighed in on Tesla's claim of invention.

Ironically, despite his efforts to establish primacy as radio's inventor, Tesla never embraced the actual device. "The radio, I know I'm its father, but I don't like it," he said. "It's a nuisance. I never listen to it. The radio is a distraction and keeps you from concentrating."[20]

Experts took different sides in the court battles. Coming to Tesla's defense was John Stone Stone, a respected research scientist for Bell Labs in Boston. Marconi, he testified in 1916, focused incorrectly on electric radiation and "it was a long time before he seemed to appreciate the real role of the earth.... Marconi's view led many to place an altogether too limited scope to the possible range of transmission." Calling Tesla's work "trail blazing," Stone declared modern radio technology had "returned to the state where Tesla developed it." Stone also offered what was becoming a frequent observation: "I think we all misunderstood Tesla. He was so far ahead of his time that the rest of us mistook him for a dreamer."[21]

Meanwhile, Columbia professor Michael Pupin stepped forward to take credit for the radio: "I invented wireless before Marconi and Tesla, and it was I who gave it unreservedly to those who followed!"[22] Acknowledging he failed to file a patent for such an important invention, Pupin then testified, "In my opinion, the first claim for wireless telegraphy belongs to Mr. Marconi absolutely, and to nobody else."[23]

Tesla could not hide his shock and outrage. "Watching his fellow Serb upon the stand," reported one newspaper, "Tesla's jaw dropped so hard, it almost cracked upon the floor."[24] When Tesla finally testified, he systematically documented his assertions for being the original inventor with transcripts from public articles and speeches as well as reports from his wireless demonstrations to large audiences in St. Louis in 1893. Tesla also compared the diagrams within his and Marconi's patent applications and declared, "You will find that absolutely not a vestige of that apparatus of Marconi remains" in contemporary radio systems.[25]

A fellow immigrant and inventor, Michael Pupin should have been Tesla's ally. Like Tesla, Pupin had grown up on a farm in Serbia's military frontier. Both of their fathers were village leaders, their grandfathers had been war heroes, and each had avoided military service. Pupin first arrived in the United States in 1874, at the age of fifteen with only a few pennies in his pocket, and he spent several tough years shoveling coal for fifty cents a ton. The two young scientists met in New York City and often discussed their early struggles, and Tesla even helped Pupin master American English in order to help him hold his job at the telephone company.

Both inventors wore their dark hair swept back and sported brush mustaches, but they displayed different personalities, different visions, and different approaches to science. As one friend of both men put it, "Telsa was an introvert, very much closed in himself and his science. Pupin was a worldly extrovert who married in wealth and believed in wealth. Tesla was a poor saintly man for whom science was his only reality. Our great sculptor Ivan Mestrovic used to tell me that Tesla was tormented and complex while Pupin was very jovial and easy to be with. But Tesla was deeper in his scientific efforts than Pupin."[26]

Pupin eventually won scholarships to the University of Berlin and Cambridge University and he returned to America in 1889 as a respected physicist who helped start Columbia College's electrical engineering program. The professor initially defended Tesla's alternating-current system, to the point he almost lost his teaching position for "eulogizing" this new technology. The competition between the immigrants developed when Pupin sided with Elihu Thomson's claim that he had developed the first and most effective AC motor. Even when Judge Townsend in 1900 declared Tesla's primacy, Pupin persisted in his criticism and increasingly aligned himself with Tesla's rivals, especially Thomson and Marconi.

The two Serbs also had a fundamental difference of opinion regarding telephone and telegraph lines. Pupin sought incremental improvements to the existing systems and Tesla wanted to replace them outright with wireless technology.

Tesla thrived on inventing and charmed patent regulators. Pupin worked primarily as a professor and suffered contentious relations with the commissioner of patents. Pupin, for instance, filed a submission in February 1894 claiming to be "the first to make a practical application of this principle in multiple telegraphy," but the unimpressed commissioner rejected Pupin's claims, arguing his ideas had been preceded by "patents by Thomson and Rice [and by] Tesla's article 'Experiments in Alternating Currents.'" The official went further to say Pupin had simply "multiplied Mr. Tesla's electric light circuits" and showed no new invention.[27]

Pupin retaliated with a four-hundred-page autobiography—*From Immigrant to Inventor*—that completely ignored Tesla's contributions to electrical science. Pupin's book won the Pulitzer Prize in 1924, bringing attention to his version of history. In expansive sections on alternating current, Pupin applauded Michael von Dobrowolsky for being the first to use a three-phase system and Elihu Thomson for developing the AC motor. Overlooking Tesla's pioneering leadership, he asserted, "If the Thomson-Houston Company had contributed nothing else than Elihu Thomson to [General Electric], it would have contributed more than enough."[28]

(Almost all other leading scientists disagreed with Pupin's oversight. Even Charles Steinmetz, the General Electric scientist who spent years trying to usurp or discount Tesla's efforts, declared, "I cannot agree with [Pupin] in the least for [a three-phase system] already existed in the old Tesla motor.... I cannot see anything new... in the new... Dobrowolsky system."[29] Bernard A. Behrend, who wrote a definitive work on the AC motor, also concluded, "The induction motor, or rotary field motor, was invented by Mr. Nikola Tesla, in 1888.")[30]

From Tesla's perspective, Pupin betrayed him during the Marconi patent trial. Tesla actually had asked Pupin "to testify on my behalf as a countryman," yet he sided with the Italian. Tesla later huffed, "Let the future tell the truth and evaluate each one according to his work and accomplishments. The present is theirs, the future, for which I really worked, is mine."[31]

The Columbia College professor even refused to offer a tribute for Tesla's seventy-fifth birthday celebration, complaining, "In the beginning of the World War a difference of opinion created a split between Mr. Tesla and myself. Neither he nor I have ever had, since that time, the opportunity to cure that split."[32] Yet when Pupin fell ill in 1935, he reached out and his Serbian colleague visited. According to one account, "Tesla approached the sick man and held his hand out and said, 'How are you, my old friend?' Pupin was speechless from emotion. He cried and tears were coming down his face."[33] Two years after the visit, Pupin died, and Tesla attended the funeral.

Another scientist who should have been close to Tesla was Charles Proteus Steinmetz. Also an immigrant, born in Prussia nine years after Tesla, he, too, was a brilliant outsider. While Tesla displayed quirks, Steinmetz simply appeared different—he was an uneven-legged, hunchback dwarf, who stood only four feet tall and whose feet, hands, and head appeared too big. Both men built towers—Tesla to send wireless transmissions, Steinmetz to attract natural lightning. Both also had the capacity to do complex mathematical calculations in their heads.

Although Steinmetz defended Tesla against some of Pupin's and Elihu Thomson's claims, he worked for thirty years at General Electric trying to obtain recognition for his own alternating-current advances. In fact, he was hired by GE in 1893 to engineer his way around Tesla's patents, and when he wrote his own history of alternating current he too virtually ignored Tesla.

The "little giant," as he was known, fled Germany in 1888 when police began harassing him for joining a socialist university group and writing articles for a local socialist newspaper. He arrived in the United States a year later. Although immigration officials initially turned

Steinmetz away because of his physical deformities, an American friend argued convincingly that his mental prowess would benefit the country. Steinmetz Americanized his first name from "Karl August Rudolph" to simply "Charles," and added a middle name of "Proteus," the shape-shifting sea god from the *Odyssey* who knew many secrets and often returned to his human form as a hunchback.

Charming and spry, often with a Blackstone panatela cigar hanging from his lips and his pince-nez glasses perched on his nose, Steinmetz regaled scientists and socialites alike. In his massive greenhouse, he collected rare plants—particularly cacti, ferns, and orchids as well as lethal creatures, including rattlesnakes, black widow spiders, and alligators.

For many years, Steinmetz was the face of General Electric, which provided him with a football-field-sized laboratory. Company ads exaggerated that he developed alternating current, that he installed the transformers and strung the wires that made the Niagara Falls project possible, and that he was "one of the men most responsible for the electrification of America."[34]

No doubt Steinmetz was responsible for innovative discoveries in hysteresis, the discipline that studies "the lag in response exhibited by a body in reacting to changes in the forces, especially magnetic forces, affecting it."[35] The scientist, for instance, calculated for how much time an iron rod maintains some of its magnetization after it has been removed from a magnetic field. Such mathematical formulae enabled engineers to improve electric motors and other electromagnetic equipment.

Like Tesla, Steinmetz relished thunderbolts, both generating his own and capturing those from nature. Unlike the reclusive Tesla, however, Steinmetz operated within a large corporate structure and understood, despite his enormous self-confidence, that inventing was collaborative.

Tesla's financial troubles continued to increase. After twenty years, the Waldorf in July 1917 asked him to leave for unpaid bills surpassing $400,000 in today's dollars. The Hotel St. Regis did the same seven years later in 1924. To satisfy a judgment, it sent a deputy sheriff to seize Tesla's laboratory furnishings, but the clever inventor talked his way out of trouble by claiming he needed his last five dollars to buy feed for hungry pigeons. The Pennsylvania Hotel also kicked him out in 1927 because the scientist's wild pigeons, which other residents referred to as "flying rats," were stinking up his room and covering the windowsills with excrement.

Pigeons, in fact, became virtually Tesla's only companions. "I have been feeding pigeons, thousands of them for years," he admitted. The inventor even arranged for hotel chefs to prepare special mixes of seeds, and he visited these winged friends regularly, as noted previously, in Bryant Park, behind the New York Public Library, often at midnight. (A corner of that park, at West 40th Street and Sixth Avenue, now features a plaque designating it as "Nikola Tesla Corner.") According to a reporter who spotted Tesla one evening sporting his derby, cane, and white gloves, "Tall, well-dressed, of dignified bearing, he whistles several times, a signal for the pigeons on the ledges of the building to flutter down about his feet. With a generous hand, the man scatters peanuts on the lawn from a bag."[36] Another journalist joined Tesla on one of those evening walks when a policeman greeted him, "Hello, Mr. Tesla," and confirmed that the aged scientist and his pigeon-feeding habits were well known in the neighborhood.[37]

What had been an innocent attraction became a delusional obsession. Tesla spent more than two thousand dollars once to construct, "using all my mechanical knowledge," a device that supported an injured bird until its bones healed. To help another pigeon with a crossed beak and unable to pick up food, he brought it back to his hotel room and fed it gently for weeks. "To care for those homeless, hungry or sick birds is the delight of my life," he said. "It is my only means of playing."[38]

Finally, one particular feathered friend became his fixation, a bizarre tale that he described to John O'Neill, his first biographer, and William Laurence, science writer for the *New York Times*. That pigeon, Tesla said, was a "beautiful bird, pure white with light grey tips on its wings; that one was different. It was a female. I had only to wish and call her and she would come flying to me. I loved that pigeon as a man loves a woman, and she loved me. As long as I had her, there was purpose to my life."[39]

A few years later, while Tesla was lying in bed in the dark, "solving problems as usual," that white bird "flew through the open window and stood on my desk. As I looked at her I knew she wanted to tell me—she was dying. And then, as I got her message, there came a light from her eyes—powerful beams of light. When that pigeon died, something went out of my life. I knew that my life's work was finished."[40]

Tesla became increasingly remote, particularly after Robert and Katharine Johnson moved to Italy in 1920. Robert had been appointed the U.S. Ambassador to Italy—with President Woodrow Wilson rewarding the writer for his large-scale relief efforts in that country after World War I. Katharine had been sick most of the previous year, having been one of the first to succumb to the great influenza epidemic that killed twenty million people worldwide; she died four years later, at the age of sixty-nine, and in a final note ordered her husband to keep in close touch with Tesla.

One old friend and a new one tried to fill the void. Long-time supporter Bernard A. Behrend convinced the Westinghouse Electric & Manufacturing Company it should be embarrassed by Tesla's financial suffering in light of the benefits he had brought to the firm. Beginning in 1934, the firm paid Tesla a piddling one hundred twenty-five dollars a month and covered his lodging at the Hotel New Yorker, payments that Behrend and Westinghouse labeled as a "consulting contract" since Tesla would have rejected charity. (For some comparison, General Electric was paying William Stanley, another electrical pioneer on hard times, a stipend of one thousand dollars a month.) The hotel at 34th Street and Eighth Avenue offered some semblance of the luxury Tesla had grown used to. Costing over twenty-two million dollars to construct, the forty-three-story structure featured an acre-sized kitchen, five restaurants, two ballrooms, and even a hospital with its own operating room. Tesla's suite, numbered 3327 (necessarily divisible by three), included two rooms—sleeping quarters and a workspace. Dark drawers, filled with papers and equipment, lined one wall, while a neat desk and chair squeezed into a tiny alcove.

The new friend was Kenneth Swezey, a boyish-faced nineteen-year-old science writer, who took the scientist on as a cause to protect him from obscurity. Almost fifty years Tesla's junior, the young man developed a special relationship that lasted until the inventor's death. Born and raised in Brooklyn, he had dropped out of secondary school and begun to write freelance science articles for various journals, eventually including *Popular Science Monthly, Saturday Evening Post, Colliers*, and *Life*.

Tesla became the focus of Swezey's efforts, and the two often spent hours in the scientist's apartment discussing inventions and ideas, with Tesla in his favorite lounging outfit of a red robe and blue slippers. They shared dinners, went to movies, and took long walks, often feeding pigeons. Swezey essentially became Tesla's assistant, publicist, confidant, and friend.

Swezey recalled Tesla regularly calling him in the middle of night, "just to tell me something on his mind." The inventor, said Swezey, "spoke nervously, with pauses, animatedly, enthusiastically. He was figuring something, working out a problem. He was comparing one theory to another, commenting, and when he felt he had arrived at the solution he was seeking, he suddenly closed the telephone." Such early-morning, one-way conversations led the writer to conclude: "Tesla lived 365 days of the year in his own world, in that fantastic combination of poetic imagination and scientific fact."[41]

The independent scientist, according to Swezey, "almost never mentioned any other friends [and] would not let me introduce him to anyone."[42] Still, the young writer became part of the informal family, befriending Tesla's nephew, Sava Kosanovic, and Robert Johnson's daughter, Agnes.

The relationship became so familiar, commented Swezey, that Tesla took to greeting him at the door stark naked. One biographer suggested Swezey was homosexual, but, other than noting the writer had never been married and had lived his entire life in the same Brooklyn apartment, there seems to be no evidence one way or the other. Swezey himself declared Tesla to be "an absolute celibate," and Kosanovic, perhaps protecting his uncle (and Yugoslavia's native son), went out of his way to affirm that observation.

Swezey clearly admired the scientist but was not blind to his faults. If someone were a friend, said the writer, "Tesla is all graciousness. Polite, enthusiastic, caustic at times—but always well intentioned—he looks out upon the world, at seventy-one, with the keen insight, health, and vigor of a man of forty-five."[43] Yet Swezey admitted Tesla during his last twenty years "could be cantankerous at times and hard to get anything out of." Still, the young journalist observed, "In the midst of it all, he often came through with periods of his old brilliance."[44]

Tesla spoke to Swezey, for example, of a future with cell phones. "When wireless is perfectly applied . . . we shall be able to communicate with one another instantly, irrespective of distance," he said. "Not only this, but through television and telephone we shall see and hear one another as perfectly as though we were face to face, despite intervening distances of thousands of miles; and the instruments through which we shall be able to do this will be amazingly simple compared with our present telephone. A man will be able to carry one in his vest pocket." Tesla even suggested one key application of the Internet by saying, "It is more than probable that the household's daily newspaper will be printed 'wirelessly' in the home during the night."[45]

The inventor also talked with automobile manufacturers about a "carriage which, left to itself, would perform a great variety of operations involving something akin to judgment."[46] That prediction came more than a hundred years before Elon Musk began manufacturing an electric car he called the Tesla and which is perfecting self-driving features, or "a great variety of operations involving something akin to judgment."

In 1931, Swezey organized an elaborate party to celebrate Tesla's seventy-fifth birthday, complete with tributes from many of the world's leading scientists and engineers. Lee de Forest, a pioneer of radio whose Audion vacuum tube made voice transmissions possible, observed, "No one so excited my youthful imagination, stimulated my inventive ambition, or served as an outstanding example of brilliant achievement in the field I was eager to enter, as did yourself." W. H. Bragg, a winner of the Nobel Prize in Physics, wrote, "I shall never forget the effect of your experiments which came first to dazzle and amaze us with their beauty and interest." Hugo Gernsback, the noted science editor, went so far as to declare: "If you mean the man who really invented, in other words, *originated* and discovered—not merely *improved* what had already been invented by others—then without a shade of doubt Nikola Tesla is the world's greatest inventor, not only at present but in all history. . . . His basic as well as revolutionary discoveries, for sheer audacity, have no equal in the annals of the intellectual world."[47]

Swezey's collection of tributes revealed Tesla's poly-sided personality. The scientist clearly inspired others, as evidenced by an executive with an electrothermic company recalling, "I began to give consideration to the various electrical methods that might be employed for melting metal, my mind at once went back to those early demonstrations by you and the electric currents which you described."[48] A Cal Tech professor, who won the Nobel Prize in 1923 for his work on cosmic rays, remembered the long-ago demonstration at Columbia College of Tesla's coil and wrote, "Since then I have done no small fraction of my research work with the aid of the principles I learned that night."[49]

One science journal editor suggested Tesla's mind "raced ahead so fast that it was just impossible [with his temperament] to get everything down on paper. That neglect turned out to be his and our loss." On a related note, an editor observed, "Tesla had an exasperating way of giving evidence to anyone who could read between the lines that he knew a whole lot more about a subject than he actually put down. He also had a perverse tendency to explain the theoretical basis of some of his inventions in terms contrary to those commonly held."[50]

Tesla used the anniversary to reflect on aging. He avowed that "man reaches his maximum power in his old age, not in middle life. Everyone should have a decade or so to sum up his life work after seventy-five."[51] He also discussed his latest idea for an ocean thermal energy conversion system, writing a paper on the opportunity entitled "On Future Motive Power."

Time magazine, learning from Swezey of the scientists' accolades, devoted a cover story in July 1931 to Tesla, whom it described as "a tall . . . eagle-headed man" looking a bit like a ghost but still alert and with a sparkle in his blue eyes.[52] The cover portrait was painted by the experienced but provocative Princess Vilna Lwoff-Parlaghy, daughter of Baroness von Zollerndorff, who shared many odd traits with Tesla. While the inventor enjoyed pigeons, for instance, she kept "two dogs, an Angora cat, a bear, lion cub, alligator, ibis, and two falcons." Also living elegantly but on the edge financially, she had recently been kicked out of the Plaza Hotel for unpaid bills.[53]

The veteran innovator, in that article, highlighted his past inventions as well as his plans to disprove Einstein's theory of relativity and to develop a new source of power, to which, he claimed, "no previous scientist has turned." He also predicted interplanetary communication "will produce a magic effect on mankind, and will form the foundation of a universal brotherhood that will last as long as humanity itself." [54] Communicating with other planets became something of an obsession, and he declared, "I would willingly sacrifice all my other achievements to realize this dream."[55]

Tesla claimed in that feature article to be "leading a secluded life" but one filled with "continuous, concentrated thought and deep meditation," which allowed him to develop a "great number of ideas." The inventor said his most serious concern "is whether my physical powers will be adequate to working them out and giving them to the world."[56]

The *Time* magazine cover story attracted the attention of another journalist, John (Jack) O'Neill, who was a correspondent for the *Herald Tribune*. The journalist became an ardent admirer, even writing gushing poems to Nikola Tesla.

> *Most glorious man of all ages*
> *Thou wert born to forecast greater days*
> *Where the wonders thy magic presages*
> *Shall alter our archaic ways.*
>
> *Your coils with their juice oscillating*
> *Sent electrical surges through the earth*
> *Sent great energy reverberating*

From the center to the outermost girth

Is the mind a power omnipresent
That fathoms the depths of all space
That speaks to an adolescent
The future triumphs of the race?[57]

Tesla thanked the young journalist "heartily" but claimed "your opinion of me is immensely exaggerated."[58] O'Neill became an acclaimed writer. He won the Pulitzer Prize for Reporting in 1937, and he wrote the first Tesla biography in 1944, entitled *Prodigal Genius*, which expanded on his glorification of Tesla as a "superman" and a "scientific saint."[59]

Despite enjoying such publicity and attention, Tesla rarely provided too much personal information. "He was a perfect gentleman, the perfect host, courtesy personified," wrote O'Neill, who spent extended hours with the scientist, "but adamant in his determination to say nothing concerning himself or his work in the presence of those who would not understand him. As a result many spectacular but false stories have been printed about him."[60]

Age seemed to have prompted more quirks. Increasingly fearful of germs, Tesla commanded the hotel staff to stay at least three feet away, and he refused to touch others, although he washed his own hands compulsively. He discarded gloves and collars after only one use. As he grew older and became more gaunt, his diet shrank, moving from meat to vegetables and then to warm milk and a concoction he called "factor actus," which consisted of white turnips, cabbage hearts, white leeks, and about ten other vegetables. He eschewed coffee, tea, and all other stimulants, but he continued to enjoy a bit of alcohol, which he described as a "veritable elixir of life." He demanded his restaurant table not be used by any other diner. According to the captain of waiters, "I had to bring him his food direct from the kitchen in an earthen pot and covered so nobody could look into it. He had to have nine napkins that he could use, one by one."[61]

Tesla overlooked his germ phobia with pigeons and family members. The chief waiter, for instance, joked about Tesla allowing the dirty birds to land all over his body and the inventor's habit of providing them with "casseroles—Germ-free." William Terbo, Tesla's grandnephew who claimed in 2014 to be the scientist's only living relative, remembered meeting the elderly inventor when he was nine. Tesla greeted the young Terbo with a traditional Serbian hug and three kisses and went out of his way to rub the lad's short hair.[62]

Finances were never Tesla's forte, and money increasingly confused him. "Money? What do I need it for?" he asked late in life. "I would enjoy it only if I had a roomful of this paper and I would throw it out the window."[63] One Westinghouse executive recalled Tesla picking up a few magazines at a newsstand, taking a "big roll of bills out of his outside overcoat pocket," handing the top bills to the newsman, and walking away without waiting for change.[64] On the few occasions he was flush, he bought expensive clothes, claiming "Money is a heavy burden to carry."[65]

According to journalist and supporter Hugo Gernsback, "The man, who, by the ignorant onlooker has been called an idle dreamer, has made a million dollars out of his inventions—and spent them as quickly on new ones. But Tesla is an idealist of the highest order and to such men, money itself means but little."[66] A virtually penniless Tesla could do little but watch companies grow hugely profitable by commercializing radio, selling lighting devices, devising robots, and developing electrotherapies—all based on his inventions.

Despite his increasing isolation from humans and modern practices, Tesla so enjoyed the media's attention he decided to host annual press conferences on each of his birthdays. According to one journalist, "Tesla, who all his life has worked in seclusion and struggles to avoid publicity with all the vigor with which movie stars court it, permits a handful of 'science writers' to violate the rules as a sort of birthday party."[67]

At least one of his remaining friends feared the press was taking advantage of Tesla, that they, in order to sell more newspapers, were portraying the inventor as an eccentric old man with wild ideas. Robert

Johnson complained, "The imaginative character of Tesla's work made him the prey of the sensational press, which ... did everything it could to exploit him for its cruel and sordid purposes, with the result of making him ridiculous only to those who had neither knowledge nor the responsibility of sober judgment."[68]

Johnson, however, did appreciate that the annual gatherings informed a broader public who "remained ignorant of the principles of which [Tesla] was a profound master and which technically were beyond their ken." As evidence of the need for such education, Johnson recalled an English lady who said to Tesla, even after he was famous for capturing the power of Niagara and sending electricity to light New York City:

"And you, Mr. Tesla, what do you do?"

"Oh, I dabble a little in electricity."

"Indeed! Keep at it, and don't be discouraged. You may end by doing something someday."[69]

At the 1932 press event, Tesla introduced his concept for a new motor powered by cosmic rays. In 1934, he outlined, with detailed diagrams, a weapon that could "send concentrated beams of [mercury] particles through the free air, of such tremendous energy that they will bring down a fleet of 10,000 enemy airplanes at a distance of two hundred fifty miles from a defending nation's border and will cause armies of millions to drop dead in their tracks."[70] The seventy-eight-year-old scientist, while drinking a quart of boiled milk, predicted his "death beam" would offer every country an "invisible Chinese wall" and make war impossible. The following year, for his seventy-ninth birthday, Tesla announced a pocket-sized but powerful oscillator that could destroy buildings. The eightieth birthday conference featured a ten-page statement, largely about cosmic rays, that also revealed his exercise regimen, which included several hundred toe wriggles each night before bed, and his prediction of living for 135 years.[71]

It was the 1933 event that offered particularly poignant insights into the aging yet insightful scientist. Aware many scoffed at his ideas, Tesla stated, "They called me crazy in 1896, when I announced the discovery of cosmic rays. Again and again they jeered when I developed something new and then, years later, saw I was right. Now, I suppose, it will be the same old story when I say I have discovered a hitherto unknown source of energy—unlimited energy that can be harnessed."[72]

Asked if he were satisfied with his accomplishments, Tesla responded, "If I were satisfied, I'd be lost. It is the challenging problems that remain that keep me going."[73]

Thomas Edison died in 1931—at the age of eighty-four and at his home "Glenmont" in West Orange, New Jersey—leaving Tesla without his one-time hero and oft-times rival. Despite their different styles, the two in their later years had maintained a cordial exchange of letters, with Tesla congratulating Edison on "the marriage of your charming daughter" and Edison often addressing his notes to "Friend Tesla."[74] In public, Tesla could be complementary, as when he applauded the Wizard of Menlo Park during the Edison Award ceremony. Yet in Edison's *New York Times* obituary, Tesla alone offered criticism, saying his competitor's "method was inefficient in the extreme ... [and he] had a veritable contempt for book learning and mathematical knowledge." Tesla called Edison "by far the most successful and, probably, the last exponent of the purely empirical method of investigation," and he reproached the great experimenter's approach by saying, "If he had a needle to find in a haystack, he would not stop to reason where it was most likely to be, but would proceed at once, with the feverish diligence of a bee, to examine straw after straw until he found the object of his search."[75] The ever-tidy Tesla even mocked Edison's slovenly habits: "If he had not later married a woman of exceptional intelligence, who made it the one object of her life to preserve him, he would have died many years ago from consequences of sheer neglect."[76]

No doubt the two experimenters took different paths. Edison ended up rumpled but rich, while Tesla became poor but dapper.[77] Yet they were among the last of the independent, cross-disciplinary, industrial age inventors, and both watched the arrival of corporate research centers and a new generation of nuclear physicists.

In what became almost a parlor game, journalists and scientists compared the two geniuses. Gernsback with *Electrical Experimenter*

argued Edison "is not so much an original inventor as a genius in perfecting existing inventions. In this respect, Tesla has perhaps been the
reverse for he has to his credit a number of brilliant as well as original inventions which, however, have not been sufficiently perfected
to permit commercial exploitation."[78] Tesla, in one of his more pride-
ful moments, defined the contrast by saying: "Edison's work on the
incandescent lamp and direct-current system of distribution was more
like the performance of an extraordinarily energetic and horse-sensed
pioneer than that of an inventor; it was prodigious in amount, but not
creative." He would praise Edison for his "vigorous pioneer work" but
Tesla complained that all the Wizard of Menlo Park "did was wrought
in known and passing forms" while he, Tesla, contributed "a new and
lasting addition to human knowledge."[79]

Tesla further suggested the distinction was between "the inventor,
who refined preexisting technology, and the discoverer who created
new principles." He put himself in the second group and Edison in the
first and argued "placing the two in the same category would completely
destroy all sense of the relative value of the two accomplishments."[80]
Edison, of course, probably would have said it differently; many scientists and journalists highlighted his proven ability to bring original discoveries to commercial fruition.

Although Edison lost the "War of the Currents," he clearly won the
struggle for public recognition. The Smithsonian's National Museum
of American History, for instance, features six large pictures of Edison
in the center of its extensive electricity exhibit, but only a small portrait of Tesla hangs on a side wall. Tesla's revolutionary AC motor is not
even included in the case of innovative devices. Sensing a plot, Samuel
Mason with the Tesla Science Foundation said Smithsonian Institution
officials told him the museum displayed little about Tesla because they
thought the scientist was not a United States citizen.[81] They told Mason
they simply viewed Edison to be "an American icon" whose prominent
images would attract more visitors.

Tesla deserves more popular attention. While Edison produced
ubiquitous consumer products such as incandescent lights and phono-

graphs, Tesla devised systems using alternating current and wireless high-frequency transmissions. No doubt those systems underpin our modern economy, even if they are little understood by those of us who benefit from them.

Increasingly withdrawn, Tesla kept up with—and sometimes commented on—current affairs. Despite his previous frustrations with J.P. Morgan, he came to that bank's defense when it fell under federal investigation. He commended the *Evening Post* for criticizing the government's actions: "The undignified character is brought into evidence more and more, and it is becoming apparent even to the dullest observer that the honor and reputation of this famous banking house is resting on a foundation as solid as the Rock of Gibraltar."[82]

While the scientist wasn't particularly political, he periodically offered his opinions. He considered prohibition, for instance, to be "a drastic, if not unconstitutional, measure."[83] Years later, he found Franklin Roosevelt's New Deal to be "destructive to established industries, and decidedly socialistic," yet he incongruously followed Roosevelt "with admiration bordering on awe," declaring the president to be "the greatest genius who appeared in the potential arena in five hundred years."[84] Shocking to today's readers, Tesla in his old age embraced eugenics and proposed "sterilizing the unfit and deliberately guiding the mating instinct. A century from now it will no more occur to a normal person to mate with a person eugenically unfit than to marry a habitual criminal."[85] The inventor's hard-to-label political views also could be surprisingly reformist, as when he said the "struggle of the human female toward sex equality will end in a new sex order, with the female as superior. . . . It is not in the shallow physical imitation of men that women will assert first their equality and later their superiority, but in the awakening of the intellect of women."[86]

As evidence of his multiple interests, Tesla also kept up with boxing, a fascination that began when he battled a bully in Lika and grew after he met John L. Sullivan, the heavyweight champion whom he referred to as a friend and "a great, baby-minded fellow, very likable."[87] Tesla claimed close relationships with Jimmy "Midland Mauler" Adamick as well as the Yugoslav Fritzie Zivic, the former welterweight boxing champion who dined with the scientist whenever he visited New York City.

Tesla, of course, approached boxing as a scientist, suggesting his "mechanical principles" allowed him to predict winners. Declaring Gene Tunney to be "at least a ten-to-one favorite" over Jack Dempsey, Tesla observed, "What counts in a contest of this kind is quickness of response. The more skillful man who can [move backward], while at the same time defending himself, can reduce greatly the intensity of his antagonist's blows while keeping his own effective. To illustrate: If a fighter attacked back with half the speed of the assailant's blow, the force of the impact will be reduced to one-half, while the energy of the impact will be reduced by one-quarter. Tunney is far superior in quickness of response."[88] (Tunney indeed won the fight, but only after the referee offered a "long count" while Tunney was floored in the seventh round.)

Year after year, reporters continued to attend the inventor's birthday press conferences, eager to fascinate readers with the eccentric genius's visions of the future. A writer with *New York World-Herald* provided some color to the 1935 event: "Twenty-odd newspapermen came away from his Hotel New Yorker birthday party yesterday, which lasted six hours, feeling hesitantly that something was wrong either with the old man's mind or else their own, for Dr. Tesla was serene in an old-fashioned Prince Albert and courtly in a way that seems to have gone out of this world."[89] As Tesla's thoughts became more prescient than practical, journalists obtained good copy but scientists increasingly mocked the aging inventor as an "idealist" with "fantastic" notions.

Reporters paid particular attention to the so-called death beams because it appeared Tesla was actually developing a high-powered gun. Inside this weapon, Tesla claimed, his magnifying transformer and a unique vacuum chamber charged minute mercury or tungsten pellets and shot them out in a narrow stream. In 1937, he wrote a treatise— *The Art of Projecting Concentrated Non-dispersive Energy through the Natural Media*—which described his "superweapon that would put an end to all war."[90] Hugo Gernsback popularized Tesla's idea in *Electrical Experimenter* by hiring the illustrator Frank Paul, considered the century's most gifted science fiction artist, to prepare cover drawings of Tesla's superweapon showering death beams onto incoming ships. (The drawings and Tesla's serialized memories helped boost the newsletter's circulation to about one hundred thousand readers.)

Tesla tried unsuccessfully to sell the gun's design to military officials in various countries, including the United States and Britain. Records suggest Tesla in 1935 received twenty-five thousand dollars (equivalent to almost four hundred fifty thousand dollars today) from the Amtorg Trading Company, an alleged Soviet front firm for developing weapons, to "supply plans, specifications, and complete information" for producing a death beam gun.[91] (Such an alleged payment is hard to reconcile with Tesla's documented poverty.) The weapon also gained the attention of a couple of thieves, or spies, who broke into his laboratory and scrutinized his papers; Tesla argued they "left empty-handed" and could learn nothing because he kept all the device's important details in his head.[92] Again, the world eventually caught up with Tesla's ingenuity, for as one media outlet several years later put it, "His death beam bears an uncanny resemblance to the charged-particle beam weapon developed by both the United States and the Soviet Union during the cold war."[93]

The annual publicity attracted an assortment of players wanting to capitalize on Tesla's fame and ideas. One was Titus deBobula, a Hungarian-born arms dealer and architect of churches, apartment buildings, and mansions. The racketeer, however, rarely paid taxes, aligned himself with pro-Hitler organizations, and was charged with selling rifle grenades and gas bombs in an effort to overthrow the Hungarian government. FBI director J. Edgar Hoover even put him on a watch list. Still, deBobula tried to befriend Tesla, commending the scientist for rejecting the Jewish Albert Einstein's relativity theories and even offering blueprints and financing to reconstruct Wardenclyffe. Tesla eventually declined the offer and "resolved to fight my own battles."[94]

As Tesla developed his "machine to end war," he reflected on the link between religion and science, the latter, he said, being "opposed to theological dogmas because science is founded on fact." Although raised by a family of priests and sometimes calling himself "deeply religious," he rejected the Christian view of an all-powerful god, suggesting instead that "the universe is simply a great machine which never came into being and never will end." To Tesla, the "soul" or "spirit" was "nothing more than the sum of the functionings of the body. When this functioning ceases, the 'soul' or the 'spirit' ceases likewise."[95] He did concede to a few superstitions, including his preference "to make important communications on Fridays and the thirteenth of each month."[96]

Tesla's one religious ritual was recognition of his Slava day, perhaps the most important annual event for a Serbian family. This partly religious and partly social festival pays tribute to one's patron saint, which for Tesla was Nicholas, a bishop in Myra in the fourth century honored for miraculously saving from death three unjustly condemned men. (Some accounts say Nicholas's saintly deed instead was to rescue three girls from prostitution by throwing three bags of gold as dowry into their window at night.) In the Julian calendar used by the Serbian Orthodox Church, Tesla's Slava day occurred on December 17, when the inventor often would dine with the Johnsons and receive greetings from George Westinghouse and other colleagues. The inventor admitted praying periodically to his patron saint, although he lamented that in his later years St. Nicholas seemed to have "forgotten me."[97]

The inventor continued to seek money . . . and be rejected. He tried to interest the U.S. government in the death beam, saying he would work on it "until I collapsed," yet he added something obviously unacceptable: "I would have to insist on one condition—I would not suffer interference from any experts. They would have to trust me."[98]

As early as 1913, not long after J. P. Morgan died, Tesla approached his son Jack but made little headway. He initially tried flattery, suggesting the young financier was "moved by the same great spirit of generosity which has animated your father, and I am more than ever desirous of enlisting your interest and support." He subsequently tried to convince the investor that he had learned to be a pragmatist, even while revealing he remained a financial romantic. "I am no longer a dreamer but a practical man of great experience gained in long and bitter trials. If I had now twenty-five thousand dollars to secure my property and make convincing demonstrations, I could acquire in a short time colossal wealth." He then made his pitch—"Would you be willing to advance me this sum if I pledged to you these inventions?" The younger Morgan declined.[99]

Despite financial frustrations, Tesla regularly expressed joy and hopefulness. In a 1935 letter to a friend, he wrote: "My life, despite some adversity, was happy and even now I have the same love for work I had in my youth."[100] On another occasion, he said, "My life has been so wonderful that it almost surpasses fiction. Every great desire I have ever had has invariably materialized."[101] Tesla acknowledged his bouts with depression but also boasted: "In the main my life is very happy, happier than any life I can conceive of."[102]

Tesla had much to be proud of. By the late 1930s, just as the inventor predicted, most American towns were wired for electricity, and families spent their evenings in bright rooms listening to radio broadcasts. They enjoyed such laborsaving motorized devices as vacuum cleaners, washing machines, stoves, toasters, irons, and hot-water heaters. To appreciate electricity's popularity, consider that in Muncie, Indiana, a midsized Midwestern town, 95 percent of homes by 1930 had electricity, even though more than a third lacked a bathtub and a fifth still relied on outhouses.

The inventor, however, was not without regrets, even if he expressed them rarely. After Robert Johnson assembled and mailed early newspaper articles about Tesla's alternating-current inventions, Tesla admitted, "The old clippings you have forwarded are a sad reminder of my former folly. I had thirty-six patents on my system of power transmission in which billions are invested now. I have won every suit without exception and had it not been for a 'scrap of paper' [the contract he tore up for George Westinghouse] I would have received in royalties Rockefeller's fortune."[103]

The financial strain led to Tesla's declining health; moreover, in 1937, at the age of eighty-one, he was hit by a taxi and thrown to the ground while on his daily walk in Manhattan. He rejected medical treatment and managed to limp home, where he virtually stayed in bed for six months. (On the night of this accident, even though suffering from shock, a wrenched back, and three broken ribs, Tesla arranged for a Postal Telegraph messenger to purchase pigeon feed and bring it to his hotel.) He never fully recovered from these injuries, and he walked thereafter with a cane.

In that same year, Tesla lost his longtime friend, Robert Johnson, who had fulfilled his wife's demand that he stay in touch with Tesla; he died at the age of eighty-four. The sequestered Tesla showed more and more signs of senility, often getting lost or forgetting addresses and events. One morning, for example, he fervently directed a messenger to deliver an envelope with money to 35 South Fifth Avenue for Mark Twain—although the address did not exist and Samuel Clemens had died years before. Growing heart problems, moreover, occasionally caused him to faint.

Tesla claimed to have been born from long-living stock, yet, recognizing his numerous youthful brushes with death and his strenuous work schedule, it's somewhat remarkable he survived into his eighties. The average life expectancy for males was then only sixty years. Noting Thomas Edison lived until eighty-four and Alexander Graham Bell until seventy-five, there might be some health benefits to an inventor's lifestyle.

A Westinghouse company representative talked with Tesla in 1939, when the inventor was eighty-three, and found him "to be thoroughly clear-headed" and his voice "sounded buoyant and enthusiastic." Although Tesla wouldn't let this assistant to a vice president upstairs, they chatted on the Hotel New Yorker line for about ten minutes and the inventor "was profuse in his thanks for my calling."[104]

Tesla spent most of that and subsequent years in his room. By 1942, he was confined to bed. He avoided former colleagues and permitted no visitors, except for summoned hotel staff. Insisting he was not ill, he refused to see a doctor.

Nikola Tesla died in his sleep in room 3327 of the Hotel New Yorker on the snowy evening of January 7, 1943, the Serbian Christmas Eve, at the age of eighty-six. A floor maid, Alice Monaghan, discovered the scientist's body the following morning, and an assistant medical examiner, Dr. Weinberg, ruled he had died suddenly of natural causes and the "police say nothing suspicious." They reported his sunken and emaciated face appeared composed, which is reflected in a death mask made later that day.

More than two thousand people attended the services at four o'clock on January 12 at the Cathedral of St. John the Divine. Episcopal Bishop William Manning agreed to hold the event but, noting the simmering conflicts between Croats and Serbs, demanded that there would be no political speeches made. For the first time in twenty years, the New York City landmark unlocked its great "golden doors" in order to accommodate the crowd. American and Yugoslav flags draped the casket, and the ceremony was conducted in Serbian by Orthodox priests. The funeral was considered an official state function by the Yugoslav government, which posted a dozen soldiers around the casket.

New York City Mayor Fiorello La Guardia offered a tribute on the radio. The president and Eleanor Roosevelt sent a letter expressing their gratitude for Tesla's contributions "to science and industry and to this country."[105] Vice President Henry Wallace wrote, "In Tesla's death the common man lost one of his best friends."[106] Three Nobel Prize winners in physics eulogized the inventor as "one of the outstanding intellects of the world who paved the way for many of the important technological developments of modern times."[107] David Sarnoff, president of RCA, said, "Tesla's mind was a human dynamo that whirled to benefit mankind."[108]

The honorary pallbearers provide a sense of the regard key scientists had for the eccentric inventor. They included Dr. Ernest F. W. Alexanderson of the General Electric Company, inventor of the Alexanderson alternator; Professor Edwin H. Armstrong of Columbia University, inventor of frequency modulation and many other important radio devices; Dr. Harvey C. Rentachler, director of the research laboratories, Westinghouse Electric & Manufacturing Company; Colonel Henry Breckenridge; Dr. Brando Cubrilovich, Yugoslav Minister of Agriculture and Supply; Consul General D. M. Stanoyavitch of Yugoslavia; and Professor William H. Barton, curator of the Hayden Planetarium.

The *New York Times'* obituary described Tesla as the "father of radio and of modern electrical generation and transmission systems."[109] John O'Neill, the award-winning science writer and Tesla's first biographer, called him "a superman who created a new world . . . [and] unquestionably one of the world's greatest geniuses."[110] Dr. Edwin Armstrong commented, "The world, I think, will wait a long time for Nikola Tesla's equal in achievement and imagination." Commenting on Tesla's place in history, Hugo Gernsback in *Electrical Experimenter* wrote, "Important as were his accomplishments, he deserves attention as a dreamer." Noting "there was no limit to the range of his imagination," the journalist predicted historians "will bracket Tesla with Da Vinci or with our own Mr. Franklin. . . . One thing is sure, the world, as we run it today, did not appreciate his peculiar greatness."[111]

That theme of being a man ahead of his time appeared again and again in the tributes. Tesla indeed had lived in flux. He was born during a lightning storm at the stroke of midnight, between today and tomorrow. He was raised a Serb into an Orthodox family in a region dominated by Croats and Roman Catholics. His father instilled religion while Tesla embraced science. This inventor craved isolation but could be a master showman. He enjoyed lavish living but walked away from lucrative contracts. He won the "War of the Currents" but died almost penniless and feeding pigeons. Tesla was one paradox after another. Maybe we all are, but Tesla's personality seemed *based* on paradox.

One science writer concluded, "He was so far ahead of his contemporaries that his patents often expired before they could be put to practical use."[112] Another calculated that Tesla made at least five outstanding scientific discoveries—cosmic rays, artificial radioactivity, disintegrating beam of electrified particles, electron microscope, and X-rays—that others "rediscovered" up to forty years later and for which they then won Nobel Prizes.[113] Speaking of Tesla's visionary work, Major General J. O. Mauborgne, former chief signal officer for the U.S. Army, stated: "Those of us who grew up with the early wireless art and are familiar with his researches and contributions to science revere his memory as the greatest genius in the early wireless field. He was so far ahead of his day in the concept of the transmission of intelligence through space that the world never fully realized that Tesla was the real inventor of wireless transmission and reception as well as many other wonderful developments."[114]

Perhaps the most balanced tribute came from the *New York Sun*: "He was an eccentric, whatever that means. A nonconformist, possibly. At any rate, he would leave his experiments and go for a time to feed the silly and inconsequential pigeons in Herald Square. He delighted in talking nonsense; or was it? Granting that he was a difficult man to deal with, and that sometimes his prediction would affront the ordinary human intelligence, here, still, was an extraordinary man of genius. He must have been. He was seeing a glimpse into that confused and mysterious frontier which divides the known and the unknown. . . . but today we do know that Tesla, the ostensibly foolish old gentleman, at times was trying with superb intelligence to find the answers. His guesses were right so often that he would be frightening. Probably we shall appreciate him better a few million years from now."[115]

Numerous scientists added their praises for Tesla's pathbreaking work on electricity, radar, and robots, while still others suggested that more of this genius's insights—on topics ranging from interstellar communication to the wireless transmission of power—would soon bear fruit, and he would be recognized as prophetic. Eight months after his death, in fact, the U.S. Supreme Court finally ruled that Tesla's patents provided the foundation for radio.

EPILOGUE

BOLDNESS OF
IGNORANCE

While Nikola Tesla died peacefully, his passing and his lack of a will sparked both international intrigue and conspiracy theories. The world in 1943 writhed in the midst of war, and the inventor's designs for death beams suggested powerful weapons that could benefit the military of any country obtaining them.

The resulting scheming was perhaps most pronounced in the Slavic countries, where Tesla had become something of a hero. According to one writer, "Every peasant is familiar with Nikola Tesla, who left Yugoslavia as an emigrant boy to become a famous scientist in America."[1] A priest recalled "that Serbs loved to buy cigarette papers on whose cover is a picture of Nikola Tesla with his name printed, and on the back of the cover this poetical verse was printed: 'Be happy all Serbs, especially those in Lika, which has given us such a Serb—genius.'"[2]

That fame placed Tesla unwittingly in the center of political debates. Perhaps his last formal meeting in the Hotel New Yorker was with King Peter II of Yugoslavia. Peter had become king in 1934, when he was only eleven, but his father's cousin served as regent. Prince Paul signed a pact with the Nazis in March 25, 1941; two days later, a British-supported coup d'état overthrew the regent and placed the seventeen-year-old Peter in charge; yet within two weeks, German-led forces invaded Yugoslavia, which surrendered on April 17, and Peter fled by climbing down a drain-pipe. In 1942, the now-eighteen-year-old exiled monarch came to the United States to plead, unsuccessfully, for aid from President Franklin Roosevelt. To garner popular support for his cause, the monarch also met with Tesla. Although shocked by the aged Tesla's ashen appearance, Peter discussed his country's plight and his hope the inventor would return and help save Yugoslavia from the Nazis. The king, in his diaries, claimed the two men wept together "for all the sorrows that had torn apart [their] homeland."[3] At least, according to the exiled Yugoslav government, Tesla responded to the king: "I am proud to be a Serb and proud to be a Yugoslav. Preserve Yugoslavia for us."[4]

(Peter fled in April 1941 to Egypt and then to London, where he married a Greek princess, his third cousin. Leading the opposition to the Nazis within Yugoslavia was Joseph Tito, who tried to finance his efforts by freezing the king's overseas assets. Peter returned briefly to his country and attempted to reassert his authority and gain access to his family's gold, but Tito in 1945 disbanded the monarchy; the United States immediately recognized the new government, although it acknowledged "with some reservations on its domestic policies." Peter fled to the United States, where he eventually got a job at the Sterling Savings & Loan Association in Los Angeles; he died in Denver in November 1970 after suffering for many years from cirrhosis of the liver.)[5]

Sava Kosanovic, Tesla's nephew (the son of his younger sister, Marica), played a key and controversial role in the inventor's final months—and the subsequent distribution of his papers. He had been a member of the Royal Yugoslav Government in 1941, fled into exile when the country was invaded, and organized Peter's meetings in 1942 with Roosevelt and Tesla. After the war, however, Kosanovic backed Joseph Tito, a rising Croat leader and Communist supporter, and he became the ambassador to the United States for the Federal People's Republic of Yugoslavia. Kosanovic convinced his ailing uncle, who had supported King Peter, to send a letter to Tito encouraging his efforts to form a socialist government.

The nephew also pushed to have himself named administrator of Tesla's estate and to have the famous inventor's papers sent to Belgrade, where Yugoslavia had proudly established a Tesla museum. In the midst of a war, however, the U.S. government expressed skepticism of Kosanovic's allegiances. The FBI and other agencies also were well aware that Tesla claimed to have developed a death beam; allegedly, this weapon could whip our armies and aircraft—something the American authorities did not want to fall into the hands of their adversaries. Tesla, moreover, had sparked concern within the FBI when he gave an innocent speech in 1922 on his scientific discoveries to the Friends of Soviet Russia. Unconfirmed rumors suggested Tesla had kept two secret laboratories (to which no journalist was invited) on Manhattan's East Side underneath the 59th Street Bridge, near Second Avenue.[6] Also troubling to the FBI, although the inventor had offered his death-beam designs to the U.S. military, was the belief that he'd sold the construction plans to a Soviet agent of the Amtorg Trading Corporation. J. Edgar Hoover went so far as to title a memo "Espionage" and wrote that he feared Kosanovic "might make certain material available to the enemy."[7]

Shortly after Tesla's death, Kosanovic, Kenneth Swezey, and George Clark, director of an RCA museum and laboratory, entered Tesla's apartment with a locksmith and the hotel's managers. They claimed to have removed only three pictures and the testimonial book Swezey had compiled for the inventor's seventy-fifth birthday. Kosanovic, however, later complained that some journals were missing from his uncle's rooms, including a large black notebook he knew Tesla kept, and he hired a lawyer to investigate and obtain control of all his uncle's possessions. The FBI, meanwhile, arrived at the apartment shortly thereafter and alleged someone had taken "valuable papers, electrical formulas, designs, etc." Of particular interest was a missing "large box or container in his room near the pigeon cages" that the inventor had told someone "contained something that could destroy an airplane in the sky."[8]

The Bureau declared the government to be "vitally interested" in preserving Tesla's documents and turned the case over to the Office of the Alien Property Custodian (OAPC), which had jurisdiction since

Kosanovic claimed control of the papers and was not a U.S. citizen. The OAPC spent weeks trying to locate and impound all of Tesla's papers and belongings, and officers even searched the various hotels where Tesla had resided. Investigators eventually seized "12 locked metal boxes, 1 steel cabinet, 35 metal cans, 5 barrels and 8 trunks."[9]

The FBI interviewed scores of Tesla acquaintances, including a hotel manager who described Tesla as "very eccentric, if not mentally deranged during the past ten years, and it is doubtful if he has created anything of value during that time."[10] Kenneth Swezey, who had spent substantial time with the aging inventor, described the potential for state secrets to be a "legend" that resulted from Tesla being a "recluse . . . who liked to talk in mystifying terms during his later years."[11]

Still, the War Policies Unit of the Department of Justice ruled the inventor's papers needed to be scrutinized by military experts. To evaluate the sensitivity of Tesla's writings, OAPC hired John Trump, director of the Massachusetts Institute of Technology's High Voltage Research Laboratory. Trump was a safe but odd choice: safe because he was a respected engineer who served as a technical aide to what is now the CIA's Office of Scientific Intelligence; odd because Tesla had mocked the effectiveness of MIT's Van de Graaff generator, which featured two thirty-foot towers and two fifteen-foot-diameter balls. Much to Trump's and MIT's dismay, Tesla contended that his own relatively tiny coil, standing only two feet tall, provided more voltage and current.

(Trump's résumé came to feature several other distinctions. He received the National Medal of Science, was elected to the National Academy of Engineering, and was the uncle of Donald Trump, the forty-fifth president of the United States. While campaigning, Donald Trump said of his relative: "I had an uncle went to MIT who is a top professor. Dr. John Trump. A genius. It's in my blood. I'm smart. Great marks. Like really smart.")[12]

John Trump initially investigated a trunk at the Hotel Governor Clinton, whose manager years earlier had accepted as payment for Tesla's debt. Tesla claimed that inside was a "device" worth ten thousand dollars, which would detonate if opened by an unauthorized person.

When Trump and the government officials approached the trunk, "the hotel manager and employees promptly left the scene." Even Trump expressed reluctance to proceed, and he tried to muster courage by looking outside and remarking on the day's pleasant weather. He slowly removed the surrounding brown paper and discovered "a handsome wooden chest bound with brass." Trump cautiously opened the lid and found, much to his relief, a "box of the type used for Wheatstone bridge resistance measurements—a common standard item found in every electric laboratory before the turn of the century."[13]

In the presence of two investigators from the OAPC and three from Naval Intelligence, the professor also spent three days at Manhattan Storage reviewing the contents of numerous boxes and trunks. Trump reported, "It is my opinion that the Tesla papers contain nothing of value for the war effort, and nothing which would be helpful to the war." He went further, saying Tesla's "thoughts and efforts during at least the past fifteen years were primarily of a speculative, philosophical, and somewhat promotional character often concerned with the production and wireless transmission of power; but did not include new, sound, workable principles or methods for realizing such results." In closing, Trump wrote, "I can therefore see no technical or military reason why further custody of the property should be retained."[14]

Despite that conclusion, John Trump asked for a top secret classification of Tesla's 1937 paper entitled *The New Art of Projecting Concentrated Non-dispersive Energy through the Natural Media*, which offered equations and schematics for how a particle-beam weapon could destroy planes and tanks. The document, which one biographer described as "written virtually as a patent application,"[15] outlined a system to propel minute mercury particles at forty-eight times the speed of sound, something Trump said was "succinctly described" but would "not enable the construction of workable combinations of generator and tube even of limited power."[16]

Trump's findings did little to retard conspiracy theories. One, according to biographer Marc Seifer, asserted: "Secret agents break into Tesla's New Yorker Hotel safe without Kosanovic knowing, remove keys to his Hotel Governor Clinton vault, and steal the death-ray prototype, substituting the equipment Trump found a week or two later."[17] Seifer also interviewed an analyst supposedly working with the Office of Strategic Services (now the Central Intelligence Agency) who claimed to have interviewed Tesla and read his unpublished papers during the scientist's final months. Angry those papers were "hauled away ... behind the Iron Curtain," this analyst said, "the conspiracy [to block access to Tesla's writings] was massive and extremely complex, going back to J. Pierpont Morgan and his wish to suppress Tesla's wireless power distribution inventions because they threatened to provide cheap or free power for the masses."[18] Leland Anderson, an engineer and writer who gushingly considered Tesla to be "a super-nova in the galaxy of the human race," argued it was common knowledge at the time of Tesla's death that the "FBI confiscated, withheld, or examined Tesla's estate."[19]

Filmmakers Joseph Sikorski and Michael Calomino wrote a screenplay, *Fragments from Olympus: The Vision of Nikola Tesla*, highlighting a mysterious plane crash over the Atlantic Ocean on January 15, 1943, which killed FBI agents Percy Foxworth and Harold Dennis Haberfeld. The previous day, according to the filmmakers, those investigators had reviewed Tesla's papers and devices and reportedly were on "a secret mission of critical importance" to brief General Dwight D. Eisenhower when their military transport fell from the sky for unexplained reasons.[20]

Perhaps the most bizarre conspiracy theory suggests that George Scherff, Tesla's longtime office manager, had settled in Germany, where his son was recruited by Adolf Hitler to return to the United States to spy on and eventually kill Tesla. This incredible tale then had George Scherff, Jr., change his name to George Herbert Walker Bush and become the forty-first president of the United States.[21]

Even though the FBI marked its Tesla file "Closed" in late 1943, the U.S. government did not release any of the scientist's belongings and notes to the Belgrade museum until the fall of 1951. Tesla's body, meanwhile, was taken after the funeral ceremony to Ferncliffe Cemetery in Ardsley, New York, where it was cremated, which is unusual in the Orthodox Church, although Tesla's favorite Serbian poet, Zmai

Iovan Iovanovich, strongly advocated for it. Tesla's ashes remained in the United States until February 1957, when they were returned to the land of his birth.

◎ ◎ ◎

We might not be able to unravel the various conspiracy theories, yet Nikola Tesla did reveal a great deal about inventing. Tesla's style certainly was unconventional, as evidenced by his envisioning a revolutionary electric motor while walking in a Budapest park reciting Goethe poems. His creative process focused on both dreaming and investigating. "After experiencing a desire to invent a particular thing," Tesla allowed an idea to "roam around in my imagination and think about the problem without any deliberate concentration." Only then would he "choose carefully the possible solutions of the problem . . . and gradually center my mind on a narrowed field of investigation."[22]

No single process or form of intelligence dictates inventiveness. As for the Edison-Tesla debate, Tesla possessed an eidetic memory, able to recall images in his mind, while Edison tended to sketch and revise. Tesla imagined prototypes, Edison tinkered. Tesla's neat office reflected his cerebral efforts, in contrast to Edison's workbench, which was littered with wires and coils he could piece together.

Tesla argued that inventing was hard work. "My belief in a law of compensation is firm," he said. "The true rewards are ever in proportion to the labor and sacrifices made."[23] Tesla, however, went further, suggesting invention could even be painful. That day quoting Goethe in a Budapest park, the inventor claimed to have wrestled from nature the vision of an alternating-current motor "against all odds, and at the peril of my existence."[24]

He maintained inventing also required patience. "The trouble with many inventors," he contended, is "they lack the willingness to work a thing out slowly and clearly and sharply in their mind." Tesla argued experimenters should not try their first idea right off. "We all make mistakes," he said, "and it is better to make them before we begin."[25]

Tesla, moreover, argued an effective inventor possessed "instinct," which he defined as "something which transcends knowledge. We have, undoubtedly, certain finer fibers that enable us to perceive truths when logical deduction, or any other willful effort of the brain, is futile."[26]

He believed the joy of inventing went beyond the accumulation of profits. "The desire that guides me in all I do," he said, "is the desire to harness the forces of nature to the service of mankind."[27]

Despite discovering totally new concepts, Tesla sometimes leaned toward a mechanistic theory of life, similar to what Descartes propounded three hundred years before. As evidence, Tesla referenced the contemporary experiments in heliotropism that "clearly [establish] the controlling power of light in lower forms of organisms."[28] "We are automata," he wrote, "entirely controlled by the forces of the medium, being tossed about like corks on the surface of the water, but mistaking the effects of the impulses from the outside for free will."[29]

Tesla may have been the last of the individual inventors. Today's science is done mostly by groups within national laboratories or giant corporations. Leland Anderson, who also assembled a comprehensive Tesla bibliography, observed that the "day of the lone pioneer and wealthy patron had passed, and for Tesla the adjustment to working with a developmental staff was untenable."[30]

John Hays Hammond, Jr., an inventor and investor who had tried to work with Tesla, lamented the demise of "curiosity-driven" research in the face of "the overly focused research setting of today's corporate and academic structure."[31] He suggested that "for Tesla scientific research was an emotional experience." Tesla himself described his exhilaration for inventing: "I do not think there is any thrill that can go through the human heart like that felt by the inventor as he sees some creation of the brain unfolding to success," he exclaimed. "Such emotions make a man forget food, sleep, friends, love, everything."[32]

Despite the shift toward institutional inventing, the United States remains enamored with its entrepreneurs and their breakthroughs. Steve Jobs and Steve Wozniak have become legends for developing the personal computer in a Palo Alto garage, and the same is true for Larry Page and Sergey Brin launching an Internet search engine while still college students and raising their initial funds from family and friends. Yet these modern entrepreneurs are known mostly for their commercial successes and their abilities to build large corporations that often incorporate or perfect the inventions of others—attributes that were not Tesla's strong suits. Still, like these innovators, Tesla struggled to think creatively, secure financing, protect patents, and market ideas. Although born more than one hundred-sixty years ago, his story remains relevant to our nation's character and future technological innovations.

In fact, we have great need today of Tesla's example of selfless out-of-the-box thinking if we are to tackle our twenty-first-century challenges. His creativity, his fresh perspectives on nature, would be particularly valuable in the electric power industry he helped create. In the century since Samuel Insull converted electricity generators into regulated monopolies, risk-averse utilities have become the chief source of pollution and are resistant to creative entrepreneurs. Tesla understood the enormous waste and pollution associated with generating electricity, and he championed efficiency. He appreciated the inexhaustible power of the sun's rays, wind, falling water, and heat from within the earth, and he consistently sought new means to capture clean energy. He appreciated the need to bring drudgery-reducing power to everyone, which now includes some two billion people around the world without electricity access. He understood the negative health consequences of burning coal and other fossil fuels.

Although Tesla, living today, might not devise the means to transmit power without wires or cost, he probably would lead a charge for sustainability and against the carbon pollution that is changing our climate. His drive for innovation and efficiency also would improve the generation and delivery of a key component of our electronic-centric economy.

It's tempting to use modern psychological labels to describe Tesla. His rituals suggest obsessive compulsive disorder (OCD). His social awkwardness and his living in his own world could be symptoms of

Asperger's syndrome. His late-in-life fixation with pigeons might sig-
nify dementia or Alzheimer's disease. His frequent withdrawals into
himself suggest schizophrenia. While such diagnoses probably have
some degree of accuracy, Tesla was more complex than any labels.
Beyond the presence of quirks and obsessions, his early story was
marked by cultural isolation, the loss of a beloved brother, and the failure to meet his anguished father's hopes. He compensated with strict
self-discipline, and that myopic focus advanced both brilliant insights
as well as odd passions.

One psychology professor argued Tesla "suffered repressed guilt
feelings associated with the untimely death of his older brother Dane
when Tesla was five years old. In the throes of the Oedipal complex
and admittedly over-attached to his mother, young Niko experienced
great trauma not only because Dane was Djouka's favorite son but also
because Niko was at that age of gaining his sexual identity." Although
Tesla's traumatic childhood experiences certainly influenced him
throughout his life, this New Age analyst went too far to suggest Tesla
sought "older brother/mother surrogates [in] strong, maturing authority figures, such as Westinghouse and Pierpont Morgan."[33]

No doubt this unconventional intellect struggled with depression. Tesla could be highly charming but then become withdrawn. He
enjoyed spurts of great energy, followed by spells of deep despondency.
Perhaps that struggle actually fueled his creativity. Tesla's mind, suggested his admirer Kenneth Swezey, had "to be somewhat unbalanced
to overcome the momentum of flighty enthusiasm or the inertia of
destructive conventionalism."[34]

Even outside of psychiatry, Tesla was hard to label. He was criticized
for being harsh, arrogant, and egotistical. As evidenced by his time in
Pittsburgh, he had a hard time working well with others. Yet a longtime assistant fondly recalled, "His genial smile and nobility of bearing
always denoted the gentlemanly characteristics that were so ingrained
in his soul."[35] Another colleague claimed he had "a magnetic personality,
but was quiet, almost shy."[36] A friend acknowledged his "distinguished
sweetness, sincerity, modesty, refinement, generosity, and force."[37] Sev-

eral reporters described him as mesmerizing and capable of conveying an enthusiasm for life and discovery. Another journalist observed "he has that supply of self-love and self-confidence that usually goes with success. And he differs from most of the men who are written and talked about in the fact that he has something to tell."[38]

Contemporaries often commented on Tesla's strict and obsessive routines. His dinners—first at the ritzy Delmonico's restaurant and later at the lavish Waldorf Astoria Hotel—began promptly at 8:10 p.m. He or an assistant would have called in his order ahead of time—often thick steaks, preferably filet mignon—and only the head waiter would be allowed to serve the meals. He usually read the afternoon paper at his regular table by the window. Tesla dined alone, except when he entertained a group of reporters or investors, and he often returned after dinner to his lab for several hours of additional work. He visited barbers three times a week for half-hour scalp massages. He never ate lunch. He took warm baths, followed by lengthy cold showers. To stimulate his brain cells, he walked regularly, often up to ten miles a day, and he would flex his toes one hundred times before going to bed in order to relieve stress. Concerned about neatness, he dressed meticulously, regularly in a black derby, silk shirt, cane, green suede high-tops, and gray suede gloves—but absolutely no jewelry.

When overwhelmed by stress, Tesla deployed "a safety device" that served as his alternative to vacations. Reflecting the era's racism, he said, "When I am all but used up, I simply do as some of the colored races, who naturally fall asleep while white folks worry." Since Tesla believed tension pushed his body to accumulate a toxic agent, he compensated by sinking "into a nearly lethargic state, which lasts for exactly half an hour." Upon awakening, Tesla claimed "a freshness of mind and ease with which I overcome obstacles that had baffled me before."[39]

It could be said that Tesla's eccentricities eclipsed his genius. Yet Tesla's various habits and quirks do not detract from what that freshness of mind envisioned. He saw nature differently, offered insights that changed our world, and created the core foundations of our modern economy. As one writer observed, "Tesla *figuratively* saw the world in ways that no one else did in part because Tesla *literally* saw the world in ways that no one else did."[40]

Tesla, perhaps because he was an outsider in so many ways, paid little heed to the scientific establishment's views about what was impossible. This interloper with a vivid imagination claimed to possess the "boldness of ignorance."[41]

Tesla's was an often misjudged brilliance. He enjoyed both practical successes as well as prophetic visions. No doubt some of the inventor's suggestions were just plain crazy, yet he proved consistently able to foresee technologies of the future. "I think we all misunderstood Tesla," wrote one contemporary scientist. "We thought he was a dreamer and visionary. He did dream and his dreams came true; he did have visions but they were of a real future, not an imaginary one."[42]

In the seventy-five years since his death, Tesla has become something of a geek's hero. This loner and polymath is admired for directing mysterious electricity to do his bidding. He's revered for offering major and disruptive insights, particularly when today's technological advances tend to be boringly incremental and devised by specialized teams within large organizations.

Like all lives, Tesla's was filled with "what ifs." What if Charles Peck had lived longer and offered Tesla the level-headed business guidance he so needed? What if J. P. Morgan had not been distracted by his mergers and provided Tesla with another contribution? What if Tesla's brother Dane had not died? What if Westinghouse had not asked to cancel Tesla's royalty payments?

Perhaps most interestingly, what if Tesla had married Anna? We know this prodigy enjoyed the company of both women and men but he believed an effective scientist needed to be isolated. "Be alone, that is the secret of invention," he declared. "Be alone, that is where ideas are born."[43] Tesla admitted falling in love with Anna as a young man, and he later exchanged regular and personal correspondence with the wife of his best friend, yet he asserted, "I never touched a woman."[44] Only near the end of his life did Tesla question his seclusion, telling one reporter: "Sometimes I feel that by not marrying, I made too great a sacrifice to my work."[45]

The inventive and unconventional Tesla has attracted his share of eccentrics. Margaret Storm, for instance, maintained that the transcripts for her biography "were received on the Tesla set, a radio-type machine invented by Tesla in 1938 for interplanetary communication." According to Storm, whose book was printed in green ink, "Tesla was a Venusian, brought to this planet as a baby in 1856 and left in a remote mountain province in what is now Yugoslavia."[46] Another biographer, Arthur Matthews, contended he and Tesla had worked together and "traveled many times to nearby planets aboard Venusian spacecraft and that Tesla, as late as 1970, was still alive, living as an extraterrestrial."[47] Yet another biographer suggested Tesla was "a wizard from another world who threw thunderbolts from the sky."[48] One New Age writer described Tesla as an "illuminati,"[49] part of a secret society allegedly pulling the levers of power and controlling world events. The terrorist responsible for releasing sarin gas in Tokyo's subway system admired Tesla and planned to steal from the Tesla Museum in Belgrade Tesla's schematics for an earthquake-provoking oscillator. Mystic healers, running expensive seminars entitled Tesla Metamorphosis, have claimed long-distance therapeutic rays could cure cancer and birth defects. Some conspiracy theorists even suggested Tesla's death beam accidentally caused the Tunguska Event on June 30, 1908, that destroyed some two thousand square kilometers of the Siberian taiga.[50]

The inventor, however, seems to be gaining increased recognition within popular culture. The U.S. Postal Service dedicated a stamp to Tesla, *Encyclopedia Britannica* ranked him among the ten most interesting historical figures, and *Life* magazine placed him among the one hundred most famous people of the last millennium. He was played by David Bowie in *The Prestige* (2006), a movie about dueling magicians, and he is the focus of *The American Side* (2016), a film that has an actor asking if Tesla invented something and the lead character responding, "Yeah, the 20th Century." He is featured in the movie *The Current War* (2019) that dramatizes the cutthroat race to determine whose electrical system would power the modern world. Tesla's name graces Nvidia's new line of advanced microchip processors. The International Electrotechnical Conference named the unit used to measure the strength of a magnetic field—the "tesla," symbol "t"—after him, only one of three Americans ever so honored. He's featured in a comic strip by Matthew Inman, creator of the popular website "TheOatmeal .com," that explains why Tesla (and not Edison) is history's greatest scientist. And Tesla is a character in the video game "Dark Void Saga," prompting one commentator to observe: "You know you'd gone into mainstream pop glory when you're in a videogame aimed at eighteen-year-old boys."[51] Perhaps most notably, Elon Musk's company used Tesla's name for its high-end electric car—distinguished for its technological mastery, sleek lines, and environmental vision—which uses a version of the AC motor Tesla designed.

Wardenclyffe on Long Island also is getting a new lease on life. The property was purchased in 1939 by the Peerless Photo Company that manufactured emulsions for photographic film and paper; in 1969, Agfa, a giant German firm, took over that operation, but production ceased in 1987 and the facility began to decay. Decades of pollution landed the site on the state's Superfund list, from which it was finally released, after substantial cleanup, in spring 2012. About a year later, a nonprofit group, originally known as Friends of Science East and now as the Tesla Science Center at Wardenclyffe, raised through crowd funding almost $1.4 million to purchase the scientist's laboratory and surrounding property, with the goal of restoring and converting it into a science museum. A documentary film entitled *Tower to the People* reports on that effort and includes an appearance by Penn Jillette of the comic team Penn & Teller. The president of the Republic of Serbia visited the site in September 2013 to dedicate a statue of Nikola Tesla. In July 2014, Elon Musk, responding to an email appeal from the nonprofit group, contributed one million dollars to the restoration effort. Hundreds of volunteers from around the world have helped clear debris, with one man referring to the site as a cathedral and another claiming it was an "honor to mow Tesla's grass."[52]

Tesla's Lika birthplace also has been rebuilt after the Ustashi, the ultranationalist Croatian group, riddled his home with bullets and toppled several structures as part of its ethnic and religious war in the 1990s against Serbs and Jews. Now the site—including Tesla's home, his father's church of St. Apostles Peter and Paul, as well as a multimedia center—is a popular tourist attraction.

Tesla's growing profile reflects in part his principled belief that technology should transcend the marketplace and that invention should not just be tied to profits. He aimed high, perhaps higher than any other inventor. He worked tirelessly to offer electric power freely to the world, to build automatons that would reduce life's drudgery, and to provide machines that could abolish war. He never accepted the status quo. Whether or not his discoveries enriched him personally, he persevered, with optimism and a love of his process. Hypersensitive and eccentric, ingenious and starry-eyed, he was, concluded *Science* magazine, "driven by inner forces which made sheer creation the most important thing in his life."[53]

APPENDIX
THE MARVEL OF
ELECTRICITY

Electrical properties had been observed for thousands of years before people knew how to generate electricity. Ancient Egyptian texts from 2750 B.C.E. described electric fish as the "Thunderers of the Nile." Thales, a brilliant Greek philosopher, wondered some 2,600 years ago at the ability of amber, a yellowish-brown and translucent resin, to attract straw when it had been rubbed against a piece of fur or wool, the same property Tesla noted when he pet his beloved cat. In fact, the Greek word for amber is "elektron."

William Gilbert, one of Queen Elizabeth's physicians, explored the connection between electricity and magnetism, and Otto von Guericke in the mid-seventeenth century constructed the first machine to generate bursts of power. During that same period, German cleric Ewald Georg von Kleist and Dutch scientist Pieter van Musschenbroek independently developed a device—named a Leyden jar after the city of Musschenbroek's experiment—that could "store" static electricity between layers of foil in a glass jar. In addition to being a key tool for scientists, that rudimentary battery allowed electricity to become something of a sideshow wonder. To bedazzle King Louis XV, Abbe Jean-Antoine Nollet assembled seven hundred friars at a monastery in Paris and somehow convinced them to join hands in a large circle, the first man gripping one electrical contact of a Leyden jar. When the last friar touched the other electrical contact, he completed the circuit, allowing the electric charge to flow through them all. Seven hundred monks simultaneously hopped into the air, and the king and his court gasped with delight.

In September 1752, Benjamin Franklin supposedly flew a kite in a gentle rain with a metal key attached to the end of the string, drawing the electrical charge within the storm cloud to flow from the kite to the key. A Swedish scientist who tried a similar experiment a year earlier died when lightning struck the rod he was holding above his head. A bit smarter, Franklin stood inside a doorway and took hold of a dry silk ribbon rather than the wet string.

An entranced public sought more sparks and crackles, but almost no one thought electricity could do anything useful. Even Franklin was "chagrined that we have been hitherto able to produce nothing in this way of use to mankind."[1] In fact, the Leyden jar's output was limited, and the discharges from static electricity and lightning came in bursts that were hard to control. A ready and steady movement of electrons is needed for a dependable power source, and a series of scientists and inventors spent decades figuring out how to generate and control that movement.

Michael Faraday in 1821, for instance, demonstrated that spinning magnets within a loop of wire generated an electric current. James Clark Maxwell, a professor at Cambridge University, tackled the problem mathematically; he proved that light was electromagnetic radiation, or electricity vibrating at a very high frequency. Joseph Henry, an American scientist and first secretary of the Smithsonian Institution, built an embryonic electric motor in 1829. Thomas Davenport, a Vermont blacksmith, devised a battery-powered machine in 1837 that drove a small printing press. Heinrich Hertz proved that electric sparks propagate electromagnetic waves into space, envisioning the potential for radio and wireless communication.

Perhaps the first practical application of this mysterious form of energy came from Humphrey Davy, an English chemist who in the early nineteenth century demonstrated an electric-arc lamp, consisting of two carbon rods separated by a thin gap; when Davy applied an electric current to one of the electrodes, a glowing arc leaped across the gap and provided light. Some sixty years later, Charles Brush installed twelve outdoor arc lamps to brighten Cleveland's downtown, but they could not capture the residential market, in part because the dangerously high electric currents had to be kept well away from people, and because the lamp's four thousand candlepower created a glare too bright for any home. Enter Tesla, Thomas Edison, and scores of other inventors and entrepreneurs.

To get a sense of what electricity and Tesla wrought, consider the first hour of your day. You wake to an electrically triggered alarm, take a shower made possible by an electric water pump and water heater, and perhaps use a blow dryer or electric shaver. You drink coffee from an electric coffee maker, which is also on a timer, pour orange juice from a carton stored in the refrigerator, and slide bread into the toaster. You check your emails on a smart phone or computer and listen to the weather report and news on a radio. You may push a button on your keys to unlock your car or in your car to lift the garage door. It was Tesla's inventions that paved the way not only for our all-electric homes but also for radio and remote control.

The MIT School of Engineering challenges its students to think "of five things you do or touch in a day that do not involve electricity in any way, were not produced using electricity, and are not related to your own body's internal uses of electricity." After a pause, the professor declares, "Nice try, but no way, you can't do it."[2]

While electricity pervades our modern society, its profound impacts are recent. Only two or three generations ago, rural residents lived in houses that relied on candles and kerosene lamps for light and on wood-burning stoves for heat and hot water. The first cooler often was a leaky chest on the back porch into which fifty-pound blocks of ice had to be hefted. Only one or two generations ago, families began to enjoy running water warmed by an electric heater. Wash still had to be run through a hand-powered wringer and clothes were hung outside to dry. Just in the 1960s did wall-mounted air conditioners make hot summer days more tolerable. Meanwhile, today's teenagers cannot imagine how their parents suffered through school without computers, electronic games, or cell phones.

Life without abundant electricity required fatiguing work. Rural residents chopped wood, which they stacked and carried inside to boil water. Starting and regulating the stove proved to be an art form, and burning wood produced unbearable temperatures in the summer. Even lighting a kerosene lamp proved to be tricky; if the wick was too high, the lamp would smoke, and it regularly had to be readjusted. Most chose to keep their residences dark after sunset. (Some two billion people—mostly rural residents in Africa, Latin America, and Asia—still lack access to electricity and its laborsaving powers.)

Contemporary fuels also limited a city's design. Only a hundred years ago, teams of horses or smoky locomotives pulled the only forms of public transportation, rendering impossible a commute from the suburbs. Stairs curtailed building heights to just a few stories. Factories and their workers were forced to locate near waterways where power wheels could be constructed. Inventors and investors recognized fortunes were to be made with a lighting and power system that eliminated these limitations.

The switch to electricity, however, was not guaranteed. Thomas Edison viewed gas companies as his "bitter enemies," and he complained that they were "keenly watching our every move and ready to pounce upon us at the slightest failure. Success meant world-wide adoption of our central-station plan. Failure meant loss of money and prestige and setting back our enterprise."[3] Gas lighting systems had evolved throughout the eighteenth and nineteenth centuries as wicks, enclosed containers, and polished reflectors continued to improve. The 1859 discovery of oil sparked a boom in kerosene lighting, displacing whale oil and volatile compounds drawn from heated coal, and investments in the gas industry had soared from $6.5 million in 1850 to $72 million in 1870. Yet Edison understood a gas system's shortcomings—each gas lamp, for instance, had to be lighted and snuffed out individually; the flame flickered and emitted small quantities of ammonia and sulfur; fumes would blacken the glass globe, as well as the interior of homes; and people often felt sick after a gaslight sucked the oxygen from a room. The innovator argued electricity offered a clearly better alternative.

Electricity's advance altered everyday routines. When the only option was cold water from a tap or outside well, many people used commercial laundries to wash their clothes or public bathhouses to clean themselves. With the arrival of electric water heaters and pumps, according to historian David Nye, "bathing became more frequent, laundering at home required less work, and washing dishes was easier." Even the accepted standards of cleanliness were altered: "People changed their clothes more often, and one bath night or one laundry day a week was no longer enough. Instead of dragging rugs outside a few times a year and beating them, the whole house could be vacuumed once a week."[4]

Electrical terms and the concept of being connected even invaded common speech in the late nineteenth and early twentieth centuries. As Nye explained, a musical performance could be "electrifying" and would "recharge" the listener. An intelligent man could usually "make the connection" and seldom "got his wires crossed." An effective organizer was "plugged in" to a "network."[5]

The growing electrification also changed a home's orientation. Families had been relatively independent when they relied on wood they chopped or kerosene they bought from a store. Yet the growing network of power cables, according to historian Thomas Schlereth, "required political and economic decisions by government and corporate enterprises" that captured additional authority through zoning ordinances, fire regulations, and building codes.[6]

An electric current results from the mysterious movement of electrons between atoms, the basic building blocks of matter. As we're taught in elementary science classes, electrons are the negatively charged particles orbiting the nucleus of an atom. Electricity basically features excited electrons that travel infinitesimal distances to displace each other in their orbits, but the current travels at nearly the speed of light, a staggering 186,000 miles per second. It does not move through hollow power lines; instead these solid, usually copper, wires enable stimulated electrons to bump into their atomic neighbors, creating a force.

One analogy is to line up a dozen pool balls, each one touching the next, in a perfectly straight line. When you hit the ball on one end the ball at the other end will almost immediately move. The other balls, in this case similar to jumping electrons, move only a little, often back and forth, but the ripple effect—in this case, the forward-moving electric current or electromagnetic force—is lightning fast.

Another comparison can be made to a tidal wave, which essentially is a vibration that moves a large amount of energy through the ocean without moving a large amount of actual water. As explained by science historian Christopher Cooper, "Electrons themselves do not have to move from one place to another in order to create an electrical current. Rather, what we observe as the flow of electrical charge is an impulse of energy that moves through a collection of electrons (and other charged particles) like a kind of vibration, passing from electron to electron."[7]

Unlike other fuels or forms of energy, we can't see electricity. We only get to observe the work it does.

We've only known how to use electricity less than one hundred fifty years. In fact, when engineers were asked to name the twentieth century's greatest accomplishment, they passed over the automobile, internal combustion engine, airplane, and computer chip in order to vote for the system that generates and distributes electricity.

Most of us share a rudimentary understanding of electricity generation and consumption. A power plant may burn coal or catch the wind, and the resulting electricity travels over wires hung on poles and threaded into our homes. We tap this current by flicking switches or plugging cords into wall outlets. We're familiar with some of the engineering terms, too. Electricity, for instance, can be measured in volts (essentially the pressure through the wires), with household electricity at 120 volts, car batteries at 12 volts, and flashlight batteries at 1.5 volts. Electricity use is measured in watts (essentially the rate at which electricity is consumed), with incandescent lightbulbs at 60 or 100 watts and microwave ovens and hair dryers at 1,000 or 1,200 watts.

We're less familiar with several other terms. "Frequency," which Tesla kept increasing, refers to the rate of an alternating current's oscillations. "Resistance," tracked in ohms, measures how something, particularly a wire, reduces the current flowing through it. "Ampere," or amp, represents the strength of the current, or the amount of charge transiting over time; one amp of current flows through a typical incandescent bulb, while a hair dryer uses about twelve amps. A "joule," having a bit more complicated definition, is "the energy dissipated as heat when an electric current of one ampere passes through a resistance of one ohm for one second."[8] "Horsepower" is a term James Watt, a Scottish engineer, devised in the early nineteenth century to compare the output of steam engines to the power of draft horses; it measures the rate at which work is done. A "kilowatt-hour" equals the amount of power consumed in an hour by ten 100-watt lightbulbs. Finally, materials that allow current to flow easily, such as copper wires, are called "conductors" and those that don't, such as cotton, are referred to as "insulators."

The basic electrical formula is $P = I \times V$. "P," as you might guess, measures the amount of power, or the work done by an electric current; it is measured in watts. "I" refers to the current, although it references the French term "intensité de courrant," and is measured in amperes or amps. "V" stands for voltage or pressure and is measured in volts. The formula essentially means power (watts) equals current (amps) times voltage (volts). As an example, a lightbulb is doing one watt of work when it is connected to a ten-volt power supply and one-tenth of an ampere of current is flowing through the lamp. That formula would read: $P = 0.1I \times 10V = 1 W$ (watt). Put another way, a 100-watt lightbulb in your house would need ten amps at ten volts.

The truth is that even physicists don't fully understand the fundamental properties of an electrical charge. At a high level, the engineering behind generating electricity usually begins with spinning—burning fuel, flowing water, or blowing wind turns a turbine. That mechanical energy then spins loops of copper wire inside a magnet (or a magnet past a coil of wire), forcing electrons along the wire to jump from one copper atom to another and create an electrical charge.

The symmetry of electricity can seem magical. Wind a wire around an iron core and then spin it between the poles of a large magnet and you have generated electricity. Reverse the process, run the current through the core, and you have produced mechanical rotation. One novelist described this balance by noting an "electrical motor is simply the generator's inverse."[9] Those motors now power our computers, air conditioners, elevators, subways, and virtually every other aspect of our modern economy. As noted previously, Tesla described electricity generation: "We wind a simple ring or iron with coils; we establish the connections to the generator, and with wonder and delight we note the effects of strange forces which we bring into play, which allow us to transform, to transmit, and direct energy at will."[10]

The power coming from such an electric generator is called alternating current (AC) since the direction of its flow alternates like a wave (specifically, a sine curve). Think of a point on a waterwheel; mapping its change of direction along a straight line would give you the same wave (half the time, the point on the wheel is moving up and the other half it's moving down). Sixty times per second according to standards in the United States and fifty times per second in Europe. AC's big advantage is that it can travel—its voltage (or pressure) can be adjusted fairly easily with a transformer, allowing generating companies to send high-voltage power efficiently over long distances, saving money and allowing the use of large motors and machines. Another transformer—the often silvery gray cylindrical boxes atop some distribution poles along your street or back alley—steps the power down to the safe voltage used by lightbulbs and most appliances in your home.

ACKNOWLEDGMENTS

Writing often is a lonely process, but this book has benefited from lots of helpful people.

Special thanks to Leona and Jerry Schecter, my literary agents who have been patient, persistent, and thoughtful. With good cheer, they consistently embraced and promoted this effort.

Mary Kay Zuravleff is a gifted and creative editor, as well as a lively writer of her own fiction. She reviewed numerous drafts, highlighted themes, and created flow.

Bruce Hathaway, a longtime friend, originally suggested Nikola Tesla as a subject for a fresh biography.

The team at W. W. Norton—particularly Starling Lawrence, a talented editor and writer who appreciates the nuances of history—has been great to work with.

Librarians at several institutions were welcoming and accommodating. Of particular note are those at the Library and Archives Division, Senator John Heinz History Center; Archives Center, National Museum of American History, Smithsonian Institution; and Manuscript Reading Room, Library of Congress.

My work hopefully builds on the efforts of numerous writers who have examined Nikola Tesla and his times, including Margaret Cheney,

Leland Anderson, Kenneth Swezey, W. Bernard Carlson, Marc Seifer, Nigel Cawthorne, John O'Neill, David Nye, Paul Israel, and Christopher Cooper—as well as researchers at the Nikola Tesla Museum (Belgrade), Tesla Memorial Society of New York, and Tesla Universe.

This book is dedicated to Kathryn Munson, who has been tolerant, encouraging, and supportive as this project moved slowly through its various phases.

NOTES

INTRODUCTION: EVERYWHERE IS ENERGY

1. Cleveland Moffitt, "A Talk with Tesla," *Atlanta Constitution*, June 7, 1896.
2. *Electrical World*, May 30, 1891.
3. E. Raverot, "Tesla's Experiments in High Frequency," *Electrical World*, March 26, 1892.
4. Joseph Wetzler, "Electric Lamps Fed from Space, and Flames That Do Not Consume," *Harper's Weekly*, July 11, 1891.
5. Nikola Tesla, *My Inventions* (Lexington, KY: Philovox, 2013), originally published in 1919, edited by David Major.
6. "Alternating Currents of High Frequency," *Electrical Review*, May 30, 1891.
7. Nikola Tesla, "High Frequency Experiments," *Electrical World*, February 21, 1891.
8. Marc J. Seifer, *Wizard: The Life and Times of Nikola Tesla* (New York: Citadel Books, 1996), 72.
9. "Alternating Currents of High Frequency," 184.
10. Wetzler, "Electric Lamps Fed from Space, and Flames That Do Not Consume."
11. "Mr. Tesla's High Frequency Experiments," *Industries*, July 24, 1891.
12. B. A. Behrend, "Edison Medal Award Speech, 1917."
13. "Nikola Tesla, 86, Prolific Inventor," *New York Times*, January 8, 1943.

CHAPTER 1: BORN BETWEEN TODAY AND TOMORROW

1. Although born at midnight, official birth records report the ninth, which is when Tesla's birthday was celebrated.
2. W. Bernard Carlson, *Tesla: Inventor of the Electrical Age* (Princeton: Princeton University Press, 2013), 18.
3. Tesla, *My Inventions*, 23.
4. Nikola Tesla, "Zmai Iovan Iovanovich," *The Century Magazine*, May 1, 1894.
5. Nikola Tesla, "Zmai Ivan Ivanovic, the Chief Servian Poet of To-day," in *Songs of Liberty and Other Poems*, ed. R. U. Johnson (New York: Century Company, 1897).
6. Tesla, "Zmai Iovan Iovanovich."
7. Dan Mrkich, "Nikola Tesla's Father—Milutin Tesla (1819–1879)," quoted in Carlson, *Tesla: Inventor of the Electrical Age*.
8. *Serbian Chronicle*, December 1929.
9. Tesla, *My Inventions*, p. 8.
10. *New York Herald*, 1893, quoted in Carlson, *Tesla: Inventor of the Electrical Age*.
11. Tesla, *My Inventions*, p. 10.
12. Tesla, "A Story of Youth Told by Age," *Smithsonian*, 1939.
13. Tesla, *My Inventions*, p. 10.
14. Ibid.
15. Ibid., 9.
16. Ibid.
17. Ibid.
18. Seifer, *Wizard*, 9.
19. Tesla, "A Story of Youth Told by Age."
20. Tesla, *My Inventions*, 9.
21. Tesla, "A Story of Youth Told by Age."
22. Ibid.
23. Ibid.
24. Ibid.
25. Nigel Cawthorne, *Tesla: The Life and Times of an Electric Messiah* (New York: Chartwell Books, 2014), 12.
26. Tesla, *My Inventions*, 8.
27. Ibid., 8–9.
28. Ibid., 23.
29. Ibid., 24.
30. Ibid., 23–24.
31. Ibid., 10.
32. Ibid., 25.
33. Ibid., 22–23.

34. Tesla, "A Story of Youth Told by Age."

35. Nikola Tesla, untitled note, April 23, 1893.

36. Tesla, *My Inventions*, 15.

37. W. K. Wisehart, "Making Your Imagination Work for You," *The American Magazine*, April 1921.

38. Tesla, *My Inventions*, 11.

39. Ibid.

40. Ibid., 33–34.

41. Ibid., 15.

42. Ibid., 14.

43. Wisehart, "Making Your Imagination Work for You."

44. Tesla, *My Inventions*, 13–14.

45. Ibid., 14.

46. Ibid., 12–13.

47. Seifer, *Wizard*, 18.

48. Tesla, *My Inventions*, 17.

49. Nikola Pribic, "Nikola Tesla: A Yugoslav Perspective," *Tesla Journal*, 6&7, 59–61.

50. Tesla, *My Inventions*, 53.

51. Ibid.

52. Ibid.

53. Ibid., 26.

54. John J. O'Neill, *Prodigal Genius: The Life of Nikola Tesla* (New York: Cosimo, 2006).

55. Tesla, *My Inventions*, 27–28.

56. Ibid., 12–13.

57. Ibid., 13.

58. Ibid., 15.

59. Nikola Tesla, "Speech on Receiving Edison Medal," Swezey Papers, 1917.

60. Tesla, *My Inventions*, 18.

61. Ibid.

62. Ibid., 19.

63. Nikola Tesla to George Seely of U.S. Patent Office, February 5, 1899.

64. Dan Mrkich: *Tesla: The European Years* (Ottawa: Commoners' Publishing, 2004), 73–74.

65. Tesla, *My Inventions*, 31.

66. Ibid.

67. Ibid., 32.

68. Nikola Tesla, "An Autobiographical Sketch," *Scientific American*, June 5, 1915.

69. Tesla, *My Inventions*, 57.

70. Ibid.

71. O'Neill, *Prodigal Genius*, 42.

72. Tesla, *My Inventions*, 56.

73. Kosta Kulishich, "Tesla Nearly Missed His Career as Inventor: College Roommate Tells," *Newark News*, August 27, 1931.

74. Seifer, *Wizard*, 17.

75. O'Neill, *Prodigal Genius*, 44.

76. Tesla, *My Inventions*, 14.

77. Wisehart, "Making Your Imagination Work for You."

78. Mrkich, *Tesla: The European Years*, 17.

79. Ibid., 76.

80. Tesla, *My Inventions*, 18.

81. Alfred O. Tate, *Edison's Open Door* (New York: Dutton, 1938), 149.

82. Tesla, *My Inventions*, 16.

83. Seifer, *Wizard*, 19.

84. Ibid., 245.

85. Tesla, "An Autobiographical Sketch."

86. Ibid.

87. Quoted in Carlson, *Tesla: Inventor of the Electrical Age*.

CHAPTER 2: A GLORIOUS DREAM

1. Edmund Morris, "Edison Illuminated: The Seventh Volume of Thomas Edison's Papers," *New York Times Sunday Book Review*, March 23, 2012.

2. Tesla, *My Inventions*, 33–34.

3. Tesla, "An Autobiographical Sketch."

4. Dragislav Petkovich, "A Visit to Nikola Tesla," *Politicka*, April 27, 1927.

5. Tesla, *My Inventions*, 60–61.

6. Ibid.

7. Ibid.

8. Ibid., 59.

9. Ibid.

10. Ibid., 41.

11. https://www.gutenberg.org/files/14591/14591-h/14591-h.htm.

12. Tesla, *My Inventions*, 61.

13. O'Neill, *Prodigal Genius*, 49.

14. Ibid.

15. Wisehart, "Making Your Imagination Work for You."

16. Alden P. Armagnac, "A Famous Prophet of Science Looks Into the Future," *Popular Science Monthly*, November 1928.

17. Seifer, *Wizard*.

18. Carlson, *Tesla: Inventor of the Electrical Age*, 54–55.
19. Tesla, *My Inventions*, 73.
20. Ibid., 35.
21. Szigeti, 1889 deposition.
22. "Nikola Tesla and His Wonderful Discoveries," *The Electrical World*, April 29, 1893.
23. 1915 Biographical Sketch, A198.
24. Ibid.
25. Tesla, *My Inventions*, 7.
26. Ibid., 65.
27. Silvanus R. Thompson, *Poly-phase Electric Currents* (New York: American Technical Book Company, 1897), 96–97.
28. "Sweeping Decision of the Tesla Patents," *Electrical Review*, September 19, 1900, 288–91.
29. Galileo Ferraris, "Electromagnetic Rotations with an Alternating Current," *Electrician*, 36 (1895), 360–75.
30. Andreas Bluhm and Louise Lippincott, *Light! The Industrial Age 1750–1900* (New York: Thames & Hudson, 2000), 31.
31. Tesla, *My Inventions*, 30.
32. Seifer, *Wizard*, 28.
33. Nikola Tesla, Motor Testimony, 189–90, 274–75.
34. Ibid., 220.
35. Tesla, *My Inventions*, 39.
36. Kenneth Swezey, "Nikola Tesla: Wonder Man of the New Wonder World," *Psychology Magazine*, October 1927.
37. Tesla, *My Inventions*, 42
38. Ibid., 70.
39. Tesla, Motor Testimony, 186.

CHAPTER 3: REVERBERATION OF HEAVEN'S ARTILLERY

1. Nikola Tesla, Note to the chairman and members of the Institute of Immigrant Welfare (undated), *Smithsonian*.
2. Tesla, *My Inventions*, 40.
3. Ibid., 41.
4. Walter Chambers, "Tesla Too Busy to Be Honored at Radio Show," September 25, 1929, Kenneth Swezey Papers, National Museum of American History.
5. Tesla, *My Inventions*, 42.
6. Tesla, Note to the chairman and members of the Institute of Immigrant Welfare.

7. Tesla, *My Inventions*, 42.

8. Ibid., 43.

9. Matthew Josephson, *Edison: A Biography* (New York: McGraw-Hill, 1959).

10. According to Wikiquote.org, spoken statement (c. 1903); published in *Harper's Monthly*, September 1932.

11. Paul Israel, *Edison: A Life of Invention* (New York: Wiley, 2000).

12. "Edison's Electric Light: The Times' Building Illuminated by Electricity," *New York Times*, September 5, 1882.

13. Robert Cornot, *A Streak of Luck* (New York: Seaview Books, 1979).

14. Israel, *Edison: A Life of Invention*.

15. Robert Silverberg, *Light for the World* (Princeton: D. Van Nostrand, 1967).

16. Tesla, *My Inventions*, 44.

17. Ibid.

18. Margaret Cheney, *Tesla: Man Out of Time* (New York: Touchstone, 1981), 55.

19. O'Neill, *Prodigal Genius*.

20. Tesla, *My Inventions*.

21. Seifer, *Wizard*, 41.

22. Tesla Electric Light and Manufacturing Company, advertisement, *Electrical Review*, September 4, 1886, 14.

23. Tesla, *My Inventions*, 44.

24. John T. Ratzlaff, ed., *Tesla Said* (Millbrae, CA: Tesla Book Company, 1984).

25. Tesla, Note to the chairman and members of the Institute of Immigrant Welfare.

26. Matthew Josephson, *Edison: A Biography* (New York: McGraw-Hill, 1959), 340.

27. Nikola Tesla, Testimony in *Complaint's Record on Final Hearing*, Vol. 1: Testimony, *Westinghouse vs. Mutual Life Insurance Co. and H. C. Mandeville* (1903) (Motor Testimony), 196.

28. Ibid., 213.

29. Nikola Tesla to Parker Page, *Smithsonian*, December 27, 1898.

30. W. A. Anthony to D. C. Jackson, March 11, 1888, quoted in Kenneth M. Swezey, "Nikola Tesla," *Science* 127 (May 16, 1958), 1149.

31. Thomas Commerford Martin, *Nikola Tesla* (1890), 106.

32. T. Commerford Martin, "Nikola Tesla," *Century*, February 1894.

33. Arthur Brisbane, "Our Foremost Electrician," *Sunday World*, July 22, 1894.

34. "Nikola Tesla and His Work," *New York Times*, September 30, 1894.

35. Martin, *Nikola Tesla*.

36. Wisehart, "Making Your Imagination Work for You."

37. "Tesla Electrifies the Whole Earth," *New York Journal*, August 4, 1897.

38. Nikola Tesla, "A New System of Alternate Current Motors and Transformers," *AIEE Transactions* 5 (September 1887–October 1888), 307–27.

39. Remarks of the Chairman, *AIEE Transactions* 5 (1887–88), 350.

40. H. R. Gardner to George Westinghouse, May 21, 1888, Heinz Center.

41. Ibid.

42. W. Stanley Jr. to George Westinghouse, June 24, 1888, in "Complainant's Record on Final Hearing, Volume II Exhibits," *Westinghouse Electrical and Manufacturing Company versus Mutual Life Insurance Company of New York and H. C. Mandeville*, U.S. Circuit Court, Western District of New York, 592–93.

43. George Westinghouse, private memorandum, July 5, 1888 (Westinghouse Corporation Archives).

44. Ibid.

45. Contract between Nikola Tesla and Westinghouse Electric Company, July 27, 1889, Heinz Center.

46. John W. Klooster, *Icons of Invention: The Makers of the Modern World from Gutenberg to Gates* (Santa Barbara, CA: Greenwood Press, 2009), 305.

CHAPTER 4: A WHIRLING FIELD OF FORCE

1. Kenneth Swezey, "Nikola Tesla," *Science*, May 16, 1958.

2. "Tesla's Split-Phase Patents," *Electrical Review*, March 22, 1899.

3. Nikola Tesla to *Electrical World*, 1914, Kenneth Swezey Papers.

4. Henry G. Proust, *A Life of George Westinghouse* (New York: Scribner's, 1926).

5. Nikola Tesla, "Death of Westinghouse," *Electrical World*, March 21, 1914.

6. Ibid.

7. Tesla, *My Inventions*.

8. Nikola Tesla, "1899 Experiments," 194.

9. O'Neill, *Prodigal Genius*, 77.

CHAPTER 5: AS REVOLUTIONARY AS GUNPOWDER WAS TO WARFARE

1. Nikola Tesla, *Nikola Tesla on His Work with Alternating Currents and their Application to Wireless Telegraphy, Telephony, and Transmission of Power: An Extended Interview*, ed. Leland Anderson (Breckenridge, CO: Twenty-first Century Books, 2002) (Referred to as Tesla, *An Extended Interview*).

2. Nikola Tesla, "Phenomena of Alternating Currents of Very High Frequency," *Electrical World* 17, February 21, 1891.

3. Tesla, *My Inventions*, 56.

4. Walter T. Stephenson, "Nikola Tesla and the Electric Light of the Future," *The Outlook*, March 9, 1895.

5. Wisehart, "Making Your Imagination Work for You."

6. Tesla, *My Inventions*, 50.

7. Thomas Commerford Martin, "Tesla's Oscillator and Other Inventions," *Century*, April 1895.

8. Ibid.

9. O'Neill, *Prodigal Genius*, 81–82.

10. Ibid., 83.

11. Nikola Tesla to Petar Mandic, August 18, 1890, in *Nikola Tesla: Correspondence with Relatives*.

12. Ibid.

13. Tesla, Motor Testimony, 235.

14. Thomas Commerford Martin, *The Inventions, Researches and Writings of Nikola Tesla* (New York: Fall River Press, 2014).

15. Leland Anderson, ed., *Nikola Tesla: On His Work with Alternating Currents and their Application to Wireless Telegraphy, Telephone, and Transmission of Power* (Denver, CO: Sun, 1992).

16. Letter from George Westinghouse to Thomas Edison dated June 7, 1888. Thomas A. Edison Papers at http://www.edison.rutgers.edu.

17. Letter from Thomas Edison to George Westinghouse dated June 12, 1888. Thomas A. Edison Papers at http://www.edison.rutgers.edu.

18. "Tesla: Master of Lighting," PBS.

19. Silverberg, *Light for the World*.

20. Jill Jonnes, *Empires of Light* (New York: Random House, 2003).

21. "Mr. Brown's Rejoiner, Electrical Dog Killing," *Electrical Engineer*, August 1888.

22. Terry S. Reynolds and Theodore Bernstein, "Edison and 'the Chair,'" *IEEE Technology & Society*, March 1989.

23. From *Electrical Review*, quoted in Cawthorne, *Tesla: The Life and Times of an Electric Messiah*, 42.

24. "Electrical Execution a Failure," *Electrical Review*, August 16, 1890, 1–2.

25. "Far Worse Than Hanging," *New York Times*, August 7, 1890.

26. Quotes from Seifer, *Wizard*.

27. "Kemmler Dies in Electric Chair," *New York Times*, August 6, 1890.

28. Martin, *The Inventions, Researches and Writings of Nikola Tesla*.

29. E. Raverot, "Tesla's Experiments in High Frequency."

30. Ibid.

31. Tesla, *An Extended Interview*, 7.

32. Ibid.

33. Milkin Radivoj to Nikola Tesla, September 24, 1895, in *Nikola Tesla: Correspondence with Relatives*, ed. and trans. Nicholas Kosanovich (1995).

34. Nikola Tesla to Petar Mandic, May 17, 1894.

35. Angelina Trbojevic to Nikola Tesla, October 9, 1898, in *Correspondence with Relatives*.

36. Marica Kosanovic to Nikola Tesla, May 11, 1902, in *Correspondence with Relatives*.

37. Angelina Trbojevic to Nikola Tesla, no date, in *Correspondence with Relatives*.

38. Anka Babic to Nikola Tesla, December 24, 1911, in *Correspondence with Relatives*.

39. Marica Kosanovic to Nikola Tesla, January 22, 1890.

40. Nikola Tesla to Petar Mandic, December 8, 1893.

41. Nikola Tesla, "My Inventions V –The Magnifying Transmitter," *Electrical Experimenter*, June 1919.

42. "Nikola Tesla's Revolution in War Telegraphy," *Philadelphia Press*, May 1, 1898.

43. "Honors to Nikola Tesla from King Alexander I," *Electrical Engineer*, February 1, 1893, 125.

CHAPTER 6: ORDER OF THE FLAMING SWORD

1. Tesla, *My Inventions*.

2. "Mr. Tesla Before the Royal Institution, London," *Electrical Review*, March 19, 1892.

3. Ibid.

4. Ibid.

5. Martin, *The Inventions, Researches and Writings of Nikola Tesla*.

6. "Mr. Tesla Before the Royal Institution, London."

7. Martin, *The Inventions, Researches and Writings of Nikola Tesla*.

8. Nikola Tesla, "Experiments with Alternative Currents of High Potential and High Frequency," *Engineering*, February 5, 1892, 171–72.

9. "Mr. Tesla's Lectures on Alternate Currents of High Potential and Frequency," *Nature*, February 11, 1892, 345.

10. "Mr. Tesla Before the Royal Institution, London," 292.

11. Ibid.

12. Ibid.

13. "Mr. Tesla's Lecture," *Electrical Review* (London), February 12, 1892.

14. Tesla, *My Inventions*, 51.

15. "Mr. Tesla and Vibratory Currents," *Electrical Engineer* (London), February 12, 1892, 157.

16. "Mr. Tesla and Rotary Currents," *Electrical Engineer* (London) January 29, 1892, 11–12.

17. "Mr. Tesla Before the Royal Institution, London."

18. A. P. Trotter, "Reminiscences," Institution of Electrical Engineers Archives (London), SC MSS 66, 532.

19. J. A. Fleming to Nikola Tesla, February 5, 1892, in Seifer, *Wizard.*

20. Tesla, *My Inventions,* 51.

21. "Mr. Tesla's Experiments of Alternating Currents of Great Frequency" (translation of Edouard Hospitalier's report in *La Nature*), *Scientific American,* March 26, 1892, 195–96.

22. "Tesla's Task of Taming Air," *Chicago Times-Herald,* May 15, 1899.

23. "Tesla's Experiments," *Electrical Review,* April 9, 1892.

24. "Mr. Tesla's Latest Motors and Transformers," *Industries,* August 22, 1890, Heinz Center.

25. Tesla, *My Inventions,* 52.

26. O'Neill, *Prodigal Genius,* 101.

27. Tesla, *My Inventions,* 75.

28. Ibid.

29. Nikola Tesla to Pajo Mandie, April 20, 1892, in *Nikola Tesla: Correspondence with Relatives* (Belgrade: Nikola Tesla Museum, 1993).

30. Nikola Tesla to J. P. Morgan, November 21, 1924, Library of Congress.

31. Tesla, *My Inventions,* 75.

32. Ibid., 76.

33. Ibid., 51–52.

CHAPTER 7: DIVINE ORGAN OF SIGHT

1. *King's Photographic Views of New York,* 1895.

2. Walter Stephenson, "Nikola Tesla and the Electric Light of the Future," *Scientific American Supplement,* March 30, 1895.

3. Hugo Gernsback, *Electrical Experimenter.*

4. Paul Israel, *Edison: A Life of Invention* (New York, Wiley, 2000).

5. Robert Silverberg, *Light for the World* (Princeton: D. Van Nostrand, 1967).

6. Starling Lawrence, *The Lightning Keeper* (New York: HarperPerennial, 2006).

7. Charles F. Scott, "Long Distance Transmission for Lighting and Power," *Electrical Engineer,* June 15, 1892.

8. Tesla, *My Inventions.*

9. T. C. Martin, "Tesla's Lecture in St. Louis," *Electrical Engineer,* March 8, 1893.

10. Martin, *The Inventions, Researches and Writings of Nikola Tesla.*

11. Ibid.

12. Arthur Brisbane, "Our Foremost Electrician," *The World,* July 22, 1894.

13. Martin, *The Inventions, Researches and Writings of Nikola Tesla.*

14. Ibid.

15. Tesla, *An Extended Interview,* 87.
16. Nikola Tesla, 1893 Lecture.
17. George Heli Guy, "Tesla, Man and Inventor," *New York Times,* March 31, 1895.
18. Matthew Josephson, *Edison: A Biography* (New York: McGraw-Hill, 1959).
19. Harold Passer, *The Electrical Manufacturers: 1875–1900* (Cambridge, MA: Harvard University Press, 1953).
20. Nikola Tesla to Henry Villard, October 10, 1892 (Houghton Library, Harvard University).
21. "World's Fair Doings," *Daily Interocean,* May 17, 1892.
22. Ibid.
23. Lewis Mumford, *Technics and Civilization* (New York: Harcourt, Brace and Company, 1934).
24. Thomas J. Schlereth, *Victorian America* (New York: HarperCollins Publishers, 1991).
25. J. P. Barrett, "Electricity," in G. R. Davis, *World's Columbian Exposition* (Chicago: Elliott Beezley, 1893).
26. Seifer, *Wizard,* 120–21.
27. "Electricians Listen in Wonder to the 'Wizard of Physics,'" *Chicago Tribune,* August 26, 1893.
28. Cheney, *Tesla: Man Out of Time,* 101–02.
29. "Mr. Tesla's Personal Exhibit at the World's Fair," *Electrical Engineer,* November 29, 1893.
30. Quote by Thomas Commerford Martin in *The Inventions, Researches and Writings of Nikola Tesla.*
31. W. Cameron, *World's Columbian Exposition* (New Haven, CT: James Brennan, 1893).
32. William Cameron, *The World's Fair: A Pictorial History of the Columbian Exposition* (New Haven, CT: James Brennan & Co., 1894).
33. Seifer, *Wizard.*
34. George Forbes, "The Electrical Transmission of Power from Niagara Falls," *Journal of the Institution of Electrical Engineers,* November 9, 1893.
35. Edward Dean Adams, *Niagara Power,* vol. 1 (Niagara Falls: Niagara Falls Power Co., 1927), 363.
36. Nikola Tesla to Edward Dean Adams, February 6, 1893.
37. Nikola Tesla to Edward Dean Adams, March 12, 1893.
38. Nikola Tesla to Edward Dean Adams, March 12 and 23, 1893.
39. Adams, *Niagara Power,* 192.
40. "What Are the Ten Greatest Inventions of Our Time?" *Scientific American,* November 1913.
41. Pierre Berton, *Niagara: A History of the Falls* (New York: Penguin, 1922).

42. C. E. L. Brown, "Reasons for the Use of the Three-Phase Current in the Lauffen Frankfurt Transmission," *Electrical World*, November 7, 1891.

43. Henry G. Prout, *A Life of George Westinghouse* (New York: Scribner's, 1926).

44. "Tesla's Work at Niagara," *New York Times*, July 16, 1895.

45. "History Making Celebration of the Only Electrical Banquet the World Has Ever Seen," *Buffalo Evening Express*, January 13, 1897.

46. Nikola Tesla, "Niagara Falls Speech," *Electrical World*, February 6, 1897.

47. Nikola Tesla, "Address on the Dedication of Niagara Falls," January 12, 1897.

48. Ibid.

49. Ibid.

50. *Buffalo Morning Express*, January 13, 1897.

51. Ibid.

52. H. G. Wells, "The End of Niagara," *Harper's Weekly*, July 11, 1906.

53. Seifer, *Wizard*, 149.

54. "Tesla's Work at Niagara," *New York Times*, July 16, 1895.

55. Charles Barnard, "Nikola Tesla, the Electrician," *The Chautauguan* 25 (1897).

56. "Tesla and Edison," *Watertower Times*, April 24, 1895.

57. Jill Jonnes, *Empires of Light* (New York: Random House, 2003).

CHAPTER 8: EARTHQUAKES AND FRIENDS

1. Forrest McDonald, *Insull* (Chicago: University of Chicago Press, 1962).

2. Burton Berry, "Mr. Samuel Insull," Insull Collection at Loyola University Chicago (unpublished, but copyrighted in 1962 by Samuel Insull, Jr.).

3. Samuel Insull's undated memo in response to questions from Mr. Martin, Insull Collection at Loyola University Chicago.

4. Seifer, *Wizard*, 79.

5. Nikola Tesla, "1893 Lecture," in *The Inventions, Researches, and Writings of Nikola Tesla*.

6. Nikola Tesla, "The True Wireless," *Electrical Experimenter*, May 1919.

7. Brisbane, "Our Foremost Electrician."

8. Nikola Tesla, "Mechanical Therapy," Anderson Collection, Heinz Center.

9. Tesla, *My Inventions*.

10. Tesla, "Mechanical Therapy."

11. F. Anderson, ed., *Mark Twain's Notebooks and Journals*, vol. 3, 1883–1891 (Berkeley, CA: University of California Press, 1979).

12. Mark Twain to Dear Mr. Tesla, November 17, 1898.

13. Earl Sparling, "Nikola Tesla, at 79, Uses Earth to Transmit Signals," *New York World-Telegram*, July 11, 1935.

14. Ibid.

15. Seifer, *Wizard*, 191.

16. Allan L. Benson, "Nikola Tesla, Dreamer," *World Today*, February 1912.

17. Cheney, *Tesla: Man Out of Time*, 152.

18. Seifer, *Wizard*.

19. Robert Johnson to Professor Osborn of Columbia University, May 17, 1894, Swezey Collection, Smithsonian.

20. Robert Underwood Johnson, *Remembered Yesterdays* (New York: Little, Brown, 1923), 401.

21. Brisbane, "Our Foremost Electrician."

22. Liet F. Jarvis Patten, "Nikola Tesla and His Work," *The Electrical World*, April 14, 1894.

23. Robert Underwood Johnson, *Songs of Liberty and Other Poems* (New York: Century, 1897).

24. Katharine Johnson to Nikola Tesla, February 8, 1898.

25. Nikola Tesla to Katharine Johnson, May 2, 1894, Library of Congress.

26. Robert Johnson to Nikola Tesla, July 28, 1896, Library of Congress.

27. Robert Johnson to Nikola Tesla, May 28, 1896, Library of Congress.

28. Robert Johnson to Nikola Tesla, December 28, 1897, Library of Congress.

29. Robert Johnson to Nikola Tesla, October 25, 1895, Library of Congress.

30. Katharine Johnson to Nikola Tesla, March 12, 1898.

31. Robert Johnson to Nikola Tesla, undated.

32. Nikola Tesla to Robert Johnson, March, 28, 1896, Library of Congress.

33. Nikola Tesla to Katharine Johnson, November 3, 1898, Library of Congress.

34. T. C. Martin to Katharine Johnson, January 8, 1894, Butler Library, Columbia University.

35. "Nikola Tesla: An Interesting Talk with America's Great Electrical Idealist," *Niagara Falls Gazette*, July 20, 1896.

36. Ibid.

37. Walter Stephenson, "Nikola Tesla and the Electric Light of the Future," *Scientific American Supplement*, March 30, 1895.

38. "How I Recharge the Battery of Life," as told by Nikola Tesla to George S. Viereck.

39. Cheney, *Tesla: Man Out of Time*, 141.

40. Seifer, *Wizard*, 123.

41. Robert Johnson to Nikola Tesla, April 9, 1899, Library of Congress.

42. Nikola Tesla to Robert Johnson, December 13, 1895.

43. Katharine Johnson to Nikola Tesla, August 6, 1896, Library of Congress.

44. Robert Johnson to Nikola Tesla, July 28, 1896, Library of Congress.

45. Seifer, *Wizard*.

46. Ibid.

47. Katharine Johnson to Nikola Tesla, February 8, 1898, Library of Congress.

48. Interview with Agnes Johnson, July 1, 1990, in Seifer, *Wizard*, 259-60.

49. Katharine Johnson to Nikola Tesla, May 3, 1896, Library of Congress.

50. Katharine Johnson to Nikola Tesla, June 6, 1898, Library of Congress.

51. Katharine Johnson to Nikola Tesla, undated, Library of Congress.

52. Katharine Johnson to Nikola Tesla, December 6, 1897.

53. Katharine Johnson to Nikola Tesla, December 20, 1903.

54. Nikola Tesla to Katharine Johnson, March 20, 1889.

55. Nikola Tesla to Katharine Johnson, February 28, 1901.

56. Nikola Tesla to Robert Johnson, December 21, 1916.

57. Nikola Tesla to Robert Johnson, December 7, 1893.

58. Nikola Tesla to Katharine Johnson, February 5, 1898.

59. Nikola Tesla to Katharine Johnson, November 3, 1898.

60. T. C. Martin to Robert Johnson, February 7, 1894, Butler Library, Columbia University.

61. Nikola Tesla to Robert Johnson, January 8, 1894.

62. Johnson, *Remembered Yesterdays*, 400.

63. Ibid., 401.

64. Robert Underwood Johnson, "In Tesla's Laboratory," http://www.teslasociety .com/underwood.htm.

65. Nikola Tesla to Katharine Johnson, March 10, 1905.

66. Nikola Tesla to Robert Johnson, December 27, 1914.

67. Nikola Tesla to Robert Johnson, January 15, 1915.

68. Curtis Brown, "A Man of the Future," *Savannah Morning News*, October 21, 1894.

69. "The Nikola Tesla Company," *Electrical Engineer*, February 13, 1895, 149.

CHAPTER 9: FIRE AND ROBOTS

1. Edward Everett Bartlett, *Edward Dean Adams* (New York: Bartlett Orr Press, 1926), 11.

2. Carlson, *Tesla: Inventor of the Electrical Age*, 209.

3. Ibid., 216.

4. Walter Stephenson, "Tesla and the Electric Light of the Future," *The Outlook*, March 9, 1895, 384.

5. "Tesla's Laboratory Burned," *Electrical Review*, March 20, 1895, 145.

6. "Mr. Tesla's Great Loss," *New York Times*, March 14, 1895.

7. T. C. Martin, "The Burning of Tesla's Laboratory," *Engineering Magazine*, April 1895.

8. *New York Tribune*, March 14, 1895.

9. Jennie Melvene Davis, "Great Master Magician Is Nikola Tesla," *Comfort*, May 1896.

10. Nikola Tesla, "Mechanical Therapy."

11. Nikola Tesla to C. F. Scott of Westinghouse Company, May 9, 1895, Library of Congress.
12. Nikola Tesla to C. F. Scott of Westinghouse Company, June 18, 1895, Library of Congress.
13. Nikola Tesla to Albert Schmidt, March 30, 1895.
14. Samuel Bannister to Nikola Tesla, April 8, 1895.
15. George Westinghouse to "Dear Mr. Tesla," August 8, 1895.
16. Nikola Tesla to George Westinghouse, June 15, 1895.
17. George Westinghouse to Nikola Tesla, June 18, 1895.
18. Nikola Tesla, "The Streams of Lenard and Roentgen and Novel Apparatus for Their Production," New York Academic of Sciences Lecture, April 6, 1897, reconstructed by Leland Anderson.
19. Nikola Tesla, "On Roentgen Rays," referenced in Cheney, *Tesla: Man Out of Time*.
20. Nikola Tesla, "High Frequency Oscillators for ElectroTherapeutic and Other Purposes," in *Proceedings of the American ElectroTherapeutic Association*, 25.
21. Seifer, *Wizard*, 169.
22. "Scotts at X-rays for the Blind," *New York Morning Journal*, December 3, 1896.
23. "Tesla Opposes Edison," *New York Evening Journal*, December 2, 1896.
24. "Edison Caught a Fluke," *New York Morning Journal*, August 10, 1897.
25. Carlson, *Tesla: Inventor of the Electrical Age*, 229.
26. "Tesla: Master of Lighting," PBS.
27. Christopher Eger, "The Robot Boat of Nikola Tesla," self-published. Referenced within Tesla profile on Wikipedia.
28. Tesla, *My Inventions*.
29. Cawthorne, *Tesla: The Life and Times of an Electric Messiah*, 77.
30. "Torpedo to Revolutionize Warfare," *New York Sun*, November 21, 1898.
31. Ibid.
32. Ibid.
33. "Tesla Declares He Will Abolish War," *New York Herald*, November 8, 1898.
34. Nikola Tesla, "The Problem of Increasing Human Energy," *The Century Illustrated Monthly Magazine*, June 1900.
35. "Tesla: Master of Lighting," PBS.
36. Carlson, *Tesla: Inventor of the Electrical Age*, 235.
37. "Tesla: Master of Lighting," PBS.
38. Ibid.
39. "Doubts Values of Tesla Discovery," *New York Herald*, November 9, 1898.
40. "Chary about Tesla's Plans," *New York Herald*, November 10, 1898.
41. "His Friends to Mr. Tesla," *Electrical Engineer*, November 24, 1898.
42. Thomas Commerford Martin to Elihu Thomson, January 16, 1917, in

H. Abrahams and M. Savin, *Selections from the Scientific Correspondence of Elihu Thomson* (Cambridge, MA: MIT Press, 1971), 352.

43. Nikola Tesla, "Mr. Tesla's Reply," *Electrical Engineer*, November 24, 1898.

44. "Tesla as 'The Wizard,'" *Chicago Tribune*, May 14, 1899.

45. "The Merrimac Destroyed," *New York Times*, June 4, 1898.

46. Martha Young, "Lieutenant Richmond P. Hobson," *Chautauguan*, 1898.

47. Carlson, *Tesla: Inventor of the Electrical Age*, 243.

48. Nikola Tesla to Robert Johnson, December 6, 1898, Library of Congress.

49. Nikola Tesla to Katharine Johnson, October 13, 1901, Library of Congress.

50. Nikola Tesla to Robert Johnson, November 8, 1898, Library of Congress.

51. Nikola Tesla, *My Inventions*, 55.

52. Nikola Tesla, "The True Wireless," *Electrical Experimenter*, May 1919.

53. Nikola Tesla, "Testimony in Patent Interference" in *Nikola Tesla: Guided Weapons & Computer Technology*, ed. Leland I. Anderson (Breckenridge, CO: Twenty-First Century Books, 1998).

CHAPTER 10: LIKE A GOD CONTROLLING NATURE'S POWERS

1. Virginia Cowles, *The Astors* (New York: Knopf, 1979), 124–25.

2. Nikola Tesla to John Jacob Astor, December 2, 1898.

3. Seifer, *Wizard*, 210.

4. John Jacob Astor to Nikola Tesla, January 18, 1899, Library of Congress.

5. "Tesla's Visit to Chicago," *Western Electrician*, May 20, 1899.

6. "Nikola Tesla Will 'Wire' to France," *Colorado Springs Evening Telegraph*, May 17, 1899.

7. Nikola Tesla to George Scherff, June 7, 1899, Library of Congress.

8. Nikola Tesla to George Scherff, September 6, 1899.

9. Nikola Tesla to George Scherff, undated.

10. George Scherff to Nikola Tesla, September 28, 1899.

11. George Scherff to Nikola Tesla, May 20, 1899.

12. George Scherff to Nikola Tesla, May 27, 1899.

13. Nikola Tesla to George Scherff, June 12, 1899.

14. George Scherff to Nikola Tesla, August 12, 1899.

15. Nikola Tesla, "The Transmission of Electric Energy Without Wires," *Electrical World and Engineer*, March 5, 1904.

16. Harry L. Goldman, "Nikola Tesla's Bold Adventure," *American West*, March 1971.

17. Carlson, *Tesla: Inventor of the Electrical Age*, 292–93.

18. *Nikola Tesla: Colorado Springs Notes, 1899–1900*, 17 (Assembled by the Nikola Tesla Museum, with an introduction and notes by Aleksandar Marincic).

19. Richard Gregg to Mrs. Nelson Hunt, October 9, 1962.

20. Tesla, "The Transmission of Electric Energy Without Wires."
21. Cawthorne, *Tesla: The Life and Times of an Electric Messiah*, 86–87.
22. O'Neill, *Prodigal Genius*, 185.
23. Tesla, "The Transmission of Electrical Energy Without Wires."
24. An example of a stationery wave occurs when someone shakes one end of a rope whose other end is attached to a wall. If the shaker adjusts her movements to be in resonance with the rope's returning vibration, she creates a single wave whose valleys and peaks seem to stand still.
25. Tesla, *My Inventions*, 56.
26. Tesla, "The Transmission of Electric Energy Without Wires."
27. Nikola Tesla, *Colorado Springs Notes, 1899–1900* (Breckenridge, CO: Twenty-First Century Books, 1978), 70.
28. George Scherff to Nikola Tesla, May 22, 1899.
29. Tesla, "The Transmission of Electric Energy Without Wires."
30. Ibid.
31. Martin, "Tesla's Oscillator and Other Inventions."
32. Tesla, "The Transmission of Electric Energy Without Wires."
33. O'Neill, *Prodigal Genius*, 183-87.
34. Nikola Tesla, "The Problem of Increasing Human Energy," *The Century Illustrated Monthly Magazine*, June 1900.
35. "Tesla Says: . . .," *New York Journal*, April 30, 1899.
36. Nikola Tesla, "The True Wireless," *Electrical Experimenter*, May 1919.
37. Nikola Tesla to Edward Whitaker (patent attorney), June 24, 1908.
38. Nikola Tesla, "System of Transmission of Electrical Energy," U.S. Patent No. 645, 576, filed on September 2, 1897 and issued on March 20, 1900.
39. "Tesla: Master of Lighting," PBS.
40. Mr. Vyvyan quoted in D. Marconi, *My Father, Marconi* (New York: McGraw-Hill, 1962).
41. "Tesla: Master of Lighting," PBS.
42. Dragislav Petrovich, "A Visit to Nikola Tesla," *Politika*, April 27, 1927.
43. Nikola Tesla to J. P. Morgan, September 5, 1902.
44. *Town Topics*, April 6, 1899, 10.
45. Nikola Tesla, "Talking with the Planets," *Collier's Weekly*, 26 (February 9, 1901), 4–5.
46. Seifer, *Wizard*, 220.
47. Colorado Springs *Gazette*, as quoted in Cheney, *Tesla: Man Out of Time*.
48. "Radio to Stars, Marconi's Hope," *New York Times*, January 19, 1919.
49. http://www.reliableplant.com/Read/27212/Edison-invention-calls-dead.
50. Robert Johnson to Nikola Tesla, undated.
51. Tesla, *Colorado Springs Notes*, October 23, 1899.
52. Tesla, *My Inventions*, 63.

53. Nikola Tesla to Robert Johnson, December 16, 1899.

54. Tesla, *Colorado Springs Notes*, January 3, 1900.

55. Ibid.

56. Ibid.

57. Carlson, *Tesla: Inventor of the Electrical Age*, 292–94.

58. Tesla, *Colorado Springs Notes*.

CHAPTER 11: SHEER AUDACITY

1. *The Electrician*, January 19, 1900, 423.

2. Nikola Tesla, "The Problem of Increasing Human Energy," *The Century Illustrated Monthly Magazine*, June 1900.

3. "Tesla: Master of Lighting," PBS.

4. Ibid.

5. Robert Johnson to Nikola Tesla, March 6, 1900, Library of Congress.

6. Nikola Tesla to Robert Johnson, undated, Library of Congress.

7. Robert Johnson to Nikola Tesla, March 6, 1900, Library of Congress.

8. Nikola Tesla to Robert Johnson, undated, Library of Congress.

9. Ibid.

10. Robert Johnson to Nikola Tesla, March 9, 1900, Library of Congress.

11. Tesla, "The Problem of Increasing Human Energy."

12. Ibid.

13. Ibid.

14. Ibid.

15. "Science and Fiction," *Popular Science Monthly* 58 (July 1900), 324–26.

16. Tesla, "The Problem of Increasing Human Energy."

17. Tesla, *An Extended Interview*, 170.

18. Nikola Tesla to George Westinghouse, April 13, 1895.

19. George Westinghouse to Nikola Tesla, December 12, 1898.

20. Nikola Tesla to George Westinghouse, January 22, 1900.

21. George Westinghouse to Nikola Tesla, September 5, 1900.

22. Nikola Tesla to George Westinghouse, January 11, 1906.

23. Francis J. Higginson to Nikola Tesla, May 11, 1899.

24. Nikola Tesla to U.S. Navy, July 11, 1899.

25. Seifer, *Wizard*, 229.

26. Nikola Tesla, "World System of Wireless Transmission of Energy," *Telegraph and Telephone Age*, October 15, 1927.

27. Carlson, *Tesla: Inventor of the Electrical Age*, 317.

28. Nikola Tesla to J. P. Morgan, February 18, 1901.

29. Nikola Tesla, "Our Future Motive Power," *Everyday Science and Mechanics*, December 1931.

30. Nikola Tesla to J. P. Morgan, February 12, 1901.

31. Lawrence Grant White to Kenneth Swezey, December 21, 1955; quoted in Seifer, *Wizard.*

32. *Nikola Tesla v. George C. Bold Jr.*, Suffolk County Supreme Court, April 1921.

33. "Tesla's Description of Long Island Plant and Inventor of the Installation as Reported in 1922 Foreclosure Appeal Proceedings," appendix 2 in Tesla, *An Extended Interview*, 191–98.

34. "Tesla's Description of Long Island Plant," 203.

35. "Cloudborn Electric Wavelets to Encircle the Globe," *New York Times*, March 27, 1904.

36. Nikola Tesla to J. P. Morgan, November 11, 1901.

37. "Wireless Signals across the Ocean," *New York Times*, December 15, 1901.

38. "Annual Dinner of the Institute at the Waldorf-Astoria, January 13, 1902, in honor of Guglielmo Marconi," *Transactions of the American Institute of Electrical Engineers*, 1902.

39. Seifer, *Wizard*, 278.

40. "T. C. Martin's Views," *New York Times*, December 15, 1901.

41. T. C. Martin to Nikola Tesla, December 17, 1901.

42. Nikola Tesla to T. C. Martin, June 3, 1903.

43. Nikola Tesla to J. P. Morgan, January 9, 1902.

44. Ibid.

45. Nikola Tesla, "Tesla Manifesto," in O'Neill, *Prodigal Genius*, 209.

46. Nikola Tesla to J. P. Morgan, January 9, 1902.

47. Nikola Tesla to J. P. Morgan, October 15, 1903.

48. Nikola Tesla to J. P. Morgan, January 9, 1902.

49. Nikola Tesla to J. P. Morgan, January 22, 1904.

50. Nikola Tesla to J. P. Morgan, January 9, 1902.

51. Nikola Tesla to J. P. Morgan, January 22, 1904.

52. Nikola Tesla to J. P. Morgan, October 13, 1904.

53. Nikola Tesla to J. P. Morgan, January 15, 1904.

54. Nikola Tesla to J. P. Morgan, July 3, 1903.

55. Nikola Tesla to J. P. Morgan, December 15, 1905.

56. Nikola Tesla, "The Transmission of Electrical Energy Without Wires as a Means for Furthering Peace," *Electrical World & Engineer*, January 7, 1905.

57. Nikola Tesla to J. P. Morgan, October 17, 1904.

58. J. P. Morgan to Nikola Tesla, June 12, 1904.

59. Nikola Tesla to J. P. Morgan, June 14, 1904.

60. J. P. Morgan office to Nikola Tesla, October 15, 1904.

61. A. V. Liebmann to Tesla Society, October 11, 1955.

62. Nikola Tesla to George Scherff, March 21, 1904.

63. Nikola Tesla to Nikola Trbojevic, undated, in *Correspondence with Relatives.*

64. Nikola Tesla to J. P. Morgan, October 13, 1904.

65. Nikola Tesla to J. P. Morgan, October 17, 1904.

66. J. P. Morgan to Nikola Tesla, December 17, 1904.

67. Seifer, *Wizard*, 300.

68. *New York World*, March 8, 1896.

69. Nikola Tesla to J. P. Morgan, December 19, 1904.

70. "Tesla's Flashes Startling," *New York Sun*, July 17, 1903.

71. Leland Anderson's notes on dinner with Muriel Arbus and Dorothy Skerritt, March 24, 1955.

72. Nikola Tesla to George Scherff, January 23, 1905.

73. "Tesla's Electrical Station Is Sold for Value of Lumber," *Colorado Springs Gazette*, June 2, 1904.

74. Nikola Tesla to Robert Johnson, January 24, 1904.

75. Nikola Tesla to J. P. Morgan, April 1, 1904.

76. Leland Anderson's notes on dinner with Muriel Arbus and Dorothy Skerritt, March 24, 1955.

77. Very Rev. Peter O. Stiyacich, *American Srbobran*, January 21, 1943.

78. *Colorado Springs Telegraph*, March 22, 1906.

79. Seifer, *Wizard*, 312.

80. George Scherff to Nikola Tesla, April 10, 1906.

81. Richmond Hobson to Miss Hull, December 22, 1903, Hobson Papers.

82. Richmond Hobson to Nikola Tesla, May 1, 1905.

83. Mrs. Richmond Hobson to Kenneth Swezey, February 14, 1956, Swezey Collection, Smithsonian Institution, National Museum of American History.

84. Ibid.

85. Ibid.

86. Tesla, *My Inventions*, 63.

87. Nikola Tesla to George Sylvester Viereck, December 17, 1934, Library of Congress.

88. "Tower to the People," a film by Joseph Sikorski.

89. Nikola Tesla to Katharine Johnson, October 16, 1907.

90. Nikola Tesla, "Sleep from Electricity," *New York Times*, October 16, 1907.

91. Nikola Tesla to George Viereck, December 17, 1934.

CHAPTER 12: TOO MUCH OF A POET AND VISIONARY

1. Cawthorne, *Tesla: The Life and Times of an Eccentric Messiah*, 126.

2. Galileo Ferraris, "Electromagnetic Rotations with an Alternating Current," *Electrician* 36 (1895), 360–75.

3. Nikola Tesla, "Can Bridge Gap to Mars," *New York Times*, June 23, 1907.

4. "Mr. Tesla Speaks Out," *New York World*, November 29, 1929.

5. Nikola Tesla to Westinghouse Company, January 29, 1930, and February 14, 1930.

6. Allan L. Benson, "Nikola Tesla, Dreamer," *Hearst's Magazine*, February 1912.

7. Tesla, *My Inventions*, 18.

8. Nikola Tesla to George Scherff, May 1, 1918, Library of Congress.

9. "Edison and Ore Refining," *IEEE Global History Network*, August 3, 2009.

10. Cheney, *Tesla: Man Out of Time*, 209.

11. Nikola Tesla, "The Future of the Wireless Art," *Wireless Telegraphy and Telephony*, 1908.

12. Nikola Tesla, "Nikola Tesla's Forecasts for 1908," *New York World*, January 6, 1908.

13. "Little Aeroplane Progress: So Says Nikola Tesla," *New York Times*, June 6, 1908.

14. Nikola Tesla, "A Lighting Machine on Novel Principles," February 7, 1918, Swezey Collection, Smithsonian Institution.

15. F. P. Stockbridge, "Tesla's New Monarch of Mechanics," *New York Herald Tribune*, October 15, 1911.

16. O'Neill, *Prodigal Genius*, 220.

17. Benson, "Nikola Tesla, Dreamer."

18. Nikola Tesla to George Scherff, May 24, 1918, Library of Congress.

19. Nikola Tesla to George Scherff, May 1, 1918, Library of Congress.

20. C. R. Possell, president of the American Development & Manufacturing Company, quoted in Seifer, *Wizard*.

21. Nikola Tesla to Charles Scott, December 30, 1908, Library of Congress.

22. *Electrical Experimenter*, August 1917, quoted in Cheney, *Tesla: Man Out of Time*, 259.

23. Cheney, *Tesla: Man Out of Time*, 265.

24. Margaret Cheney, Robert Uth, and Jim Glenn, *Tesla, Master of Lightning* (New York: Barnes & Noble Publishing, 1999), 129.

25. Nikola Tesla, "Tesla Patent 1,655,114 Apparatus for Aerial Transportation," United States Patent Office.

26. John Hammond, Jr. to Nikola Tesla, November 10, 1910, Library of Congress.

27. Nancy Rubin, *John Hays Hammond, Jr.: A Renaissance Man in the Twentieth Century* (Gloucester, MA: Hammond Museum, 1987).

28. John Hays Hammond, Jr. to Kenneth Swezey, October 26, 1956, Heinz Center.

29. Nikola Tesla to John Hays Hammond, Jr., November 12, 1910.

30. John Hays Hammond, Jr., "The Future of Wireless," *National Press Reporter*, May 1912.

31. Seifer, *Wizard*, 348.

32. Harris Hammond to Nikola Tesla, June 10, 1913.

33. Seifer, *Wizard*, 353.

34. Waldemar Kaempffert, quoted in Seifer, *Wizard*, 352.

35. Clarence Lawrence to Kenneth Swezey, January 25, 1957.

36. O'Neill, *Prodigal Genius*, 275.

37. Nikola Tesla to J. P. Morgan, February 17, 1905, Library of Congress.

38. Nikola Tesla to George Scherff, undated, Library of Congress.

39. George Scherff to Nikola Tesla, February 14, 1914.

40. Leland Anderson notes on dinner with Muriel Arbus and Dorothy Skerritt, March 24, 1955.

41. George Scherff to Nikola Tesla, June 8, 1915.

42. George Scherff to Nikola Tesla, undated, Library of Congress.

43. Nikola Tesla to George Scherff, March 26, 1909.

44. Nikola Tesla to George Scherff, January 11, 1909.

45. George Scherff to Nikola Tesla, July 30, 1906.

46. George Scherff to Nikola Tesla, December 31, 1906.

47. "Tesla at 75," *Time*, June 20, 1931.

48. Nikola Tesla, "Nikola Tesla—New York" (undated typed, 10-page manuscript) Smithsonian.

49. Michael Mok, "Nikola Tesla Wilts Fourteen Reporters with Fabulous New Science Theories," *New York Post*, July 11, 1935.

50. Albert Einstein to Nikola Tesla, June 1931.

51. T. C. Martin, "The Tesla Lecture in St. Louis," *Electrical Engineer*, March 18, 1893.

52. "Tesla Has Only Credit," *New York Times*, May 8, 1916.

53. Seifer, *Wizard*, 382.

54. Cawthorne, *Tesla: The Life and Times of an Electric Messiah*, 111.

55. Robert Johnson to Nikola Tesla, March 1, 1916.

56. Nikola Tesla to Robert Johnson, December 24, 1914.

57. Nikola Tesla to Robert Johnson, December 27, 1914.

58. Robert Johnson to Nikola Tesla, April 2, 1916.

59. "Electrified Schoolroom to Brighten Dull Pupils," *New York Times*, August 18, 1912.

CHAPTER 13: SO FAR AHEAD OF HIS TIME

1. "Tesla's Discovery Nobel Prize Winner," *New York Times*, November 7, 1915.

2. Cheney, *Tesla: Man Out of Time*, 245.

3. Anderson, ed., *Nikola Tesla: On His Work in Alternating Currents*, 48.

4. Nikola Tesla to Robert Johnson, November 10, 1915.

5. O'Neill, *Prodigal Genius*, 230.

6. Comment by Richard Sogge, as told to Leland Anderson and referenced by Carlson, *Tesla: Inventor of the Electrical Age*, 240.

7. *Minutes, Edison Medal Meeting*, American Institute of Electrical Engineers, May 18, 1917, Smithsonian Institution.
8. Ibid.
9. Ibid.
10. Nikola Tesla, "Speech on Receiving Edison Medal," 1917, Swezey Papers.
11. Ibid.
12. Nikola Tesla, "Edison Medal Speech," May 18, 1917.
13. Waltham full-page ad in *Hearst's International*, December 1922.
14. Tesla, *My Inventions*, 7.
15. Cheney, *Tesla: Man Out of Time*, 140.
16. Nikola Tesla to Hugo Gernsback, November 30, 1901.
17. Nikola Tesla to Hugo Gernsback, November 4, 1902.
18. Franklin D. Roosevelt to Navy Department, September 14, 1916.
19. W. Jolly, *Marconi* (New York: Stein & Day, 1972).
20. "Nikola Tesla's Death," *Brooklyn Eagle*, July 8, 1943.
21. Leland Anderson, ed. "John Stone Stone on Nikola Tesla's Priority in Radio and Continuous-Wave Radiofrequency Apparatus," *The Antique Wireless Review* 1 (1896).
22. "Prof. Pupin Now Claims Wireless His Invention," *Los Angeles Examiner*, May 13, 1915.
23. "When Powerful High-Frequency Electrical Generators Replace the Spark-Gap," *New York Times*, October 6, 1912.
24. Cawthorne, *Tesla: The Life and Times of an Electric Messiah*, 114.
25. "Marconi Wireless vs. Atlantic Communications Co.," 1915.
26. Bogdan Radica to Margaret Cheney (undated).
27. Pupin papers, May 21, 1895, Butler Library as noted in Seifer, *Wizard*.
28. Michael Pupin, *From Immigrant to Inventor* (New York: Scribner's, 1925), 289.
29. Charles Steinmetz, "Pupin on Polyphasal Generators," *AIEE Transactions*, December 16, 1891.
30. B. A. Behrend, *Induction Motor* (New York: McGraw Hill, 1921).
31. Cheney, *Tesla: Man Out of Time*.
32. Michael Pupin to Kenneth Swezey, May 29, 1931.
33. Stanko Stoilkovic, "Portrait of a Person, a Creator, and a Friend," *The Tesla Journal*, 1986–87.
34. Advertisement within the Leland Anderson collection.
35. http://www.dictionary.com/browse/hysteresis.
36. *New York World*.
37. Dragislav Petkovich, "A Visit to Nikola Tesla," *Politika*, April 27, 1927.
38. Cawthorne, *Tesla: The Life and Times of an Electric Messiah*, 142.
39. O'Neill, *Prodigal Genius*.
40. Ibid.

41. "100 Years Since the Birth of Nikola Tesla," *Politika*, July 8, 1956.
42. Kenneth Swezey to Mrs. Richmond Hobson, October 9, 1956, Swezey Collection, National Museum of American History.
43. Kenneth Swezey, "Nikola Tesla: Wonder Man of the New Wonder World," *Psychology Magazine*, October 1927.
44. Kenneth Swezey to Edward Hewitt, July 5, 1956.
45. John B. Kennedy, "When Woman Is Boss," *Collier's*, January 30, 1926.
46. Tesla, *My Inventions*, 81.
47. Cheney, *Tesla: Man Out of Time*, 293–94.
48. E. F. Northrup to Nikola Tesla, July 1931.
49. Robert Millikan to Nikola Tesla, July 1931, Swezey Collection.
50. Kenneth Swezey to Leland Anderson, June 28, 1955.
51. "I Expect to Talk with Mars," as told by Nikola Tesla to George Viereck (undated), Heinz Center.
52. "Tesla at 75," *Time*, June 20, 1931, 27–30.
53. Leland Anderson, "Tesla Portrait by the Princess Vilma Lwoff-Parlaghy," *The Tesla Journal*, nos. 4/51986/87.
54. "Tesla at 75," 30.
55. "I Expect to Talk with Mars."
56. "Tesla at 75."
57. John O'Neill to Nikola Tesla, February 23, 1916.
58. Nikola Tesla to J. O'Neill, February 26, 1916.
59. O'Neill, *Prodigal Genius*.
60. John J. O'Neill, "Tesla: Man of 'Inspired' Discoveries," *New York Herald Tribune*, January 24, 1943.
61. Carol Taylor, "Guests Hotelmen Never Forget," *New York World-Telegram*, January 26, 1966.
62. "Tower to the People," a film by Joseph Sikorski.
63. "100 Years Since the Birth of Nikola Tesla," *Politika* (translated from Serbo-Croatian), July 8, 1956.
64. E. H. Sniffen to the Westinghouse Company, January 3, 1939.
65. Millionaire (Nikola Tesla) to Luka (Robert Johnson), June 26, 1900.
66. Cawthorne, *Tesla: The Life and Times of an Electric Messiah*, 135.
67. John J. O'Neill, "Tesla Cosmic Ray Motor May Transmit Power Round Earth," *The Eagle*, June 1932.
68. Johnson, *Remembered Yesterdays*, 401.
69. Ibid.
70. "Tesla at 78 Bares New Death-Beam," *New York Times*, July 11, 1934.
71. Carlson, *Tesla: Inventor of the Electrical Age*, 380.
72. William Eagle, "Tesla at 77 Hopes World Soon Will Call Him Crazy—It Will

Mean Success to His New Energy Source," *New York World-Telegraph*, July 10, 1933.

73. Ibid.

74. Thomas Edison to Nikola Tesla, June 3, 1896, Swezey Collection.

75. "Tesla Says Edison Was an Empiricist," *New York Times*, October 19, 1931.

76. Ibid.

77. Observation by Margaret Cheney in *Tesla: Man Out of Time*.

78. H. Gernsback, "Edison and Tesla," *The Electrical Experimenter*, December 1915.

79. "Mr. T Speaks Out," *The World*, September 9, 1929.

80. O'Neill, *Prodigal Genius*.

81. "Tower to the People," a film by Joseph Sikorski.

82. Nikola Tesla, Letter to the Editor, *Evening Post*, June 5, 1933.

83. Tesla, *My Inventions*, 17.

84. Nikola Tesla, note to National Committee of Independent Voters, November 3, 1940.

85. Seifer, *Wizard*.

86. Kennedy, "When Woman Is Boss."

87. "Dr. Tesla Picks Tunney on Basis of Mechanics," *New York Herald Tribune*, September 22, 1927.

88. Ibid.

89. Quoted in Cheney, *Tesla: Inventor of the Electrical Age*, 380; and Cheney and Uth, *Master of Lightning*, 151–52.

90. Nikola Tesla, *The Art of Projecting Concentrated Non-dispersive Energy through Natural Media*, circa May 16, 1935.

91. Exhibits Q and D, John Trump to Walter Gorsuch, January 30, 1943.

92. John O'Neill, "Tesla Tries to Prevent World War II" (unpublished Chapter 34 of *Prodigal Genius*), PBS.

93. "Tesla: Master of Lighting," PBS.

94. Nikola Tesla to Titus deBobula, May 31, 1911.

95. Nikola Tesla, as told to George Sylvester Viereck (February 1937), *A Machine to End War*, PBS.

96. Ibid.

97. Nikola Tesla to J. P. Morgan, December 14, 1904.

98. "Nikola Tesla Dies; Prolific Inventor," *New York Times*, January 9, 1943.

99. Nikola Tesla to J. P. Morgan Jr., November 29, 1934.

100. Nikola Tesla to Florence E. Enderle, July 18, 1935, Swezey Collection.

101. Richard Atcheson, "Monument Here Will Honor an Inventor They Called Crazy," *Chicago Daily News*, September 3, 1960.

102. Cheney, *Tesla: Man Out of Time*, 141.

103. Nikola Tesla to Robert Johnson, December 24, 1914.

104. E. H. Sniffen to Westinghouse Company, January 3, 1939.

105. "2,000 Are Present at Tesla Funeral," *New York Times*, January 13, 1943.

106. Ibid.

107. Cheney, *Tesla: Man Out of Time*, 326.

108. Cawthorne, *Tesla: The Life and Times of an Electric Messiah*, 174.

109. "2,000 Are Present at Tesla Funeral."

110. John J. O'Neill, *Prodigal Genius: The Life of Nikola Tesla* (Ives Washburn, 1944).

111. Hugo Gernsback, "The Drama of Mr. Tesla," *World*, January 17, 1943.

112. Hugo Gernsback to Kenneth Swezey, June 18, 1956.

113. O'Neill, *Prodigal Genius*, 154.

114. Major General J. O. Mauborgne telegram to Kenneth Swezey, January 11, 1943.

115. "Nikola Tesla Dead," *New York Sun*, January 1943.

EPILOGUE: BOLDNESS OF IGNORANCE

1. Joel Martin Halpern, *A Serbian Village* (Boulder: Colorado University Press, 1958).

2. Very Rev. Peter O. Stiyachich, *American Srbobran*, January 1, 1943, Heinz Center.

3. Seifer, *Wizard*, 448.

4. Press release from Royal Yugoslav Government Information Center, January 11, 1943, Swezey Collection.

5. "The Sad Life of Peter II," *Chicago* magazine, January 2013.

6. Seifer, *Wizard*, 457.

7. Ibid., 448.

8. Charles Houslin to Leland Anderson, April, 12, 1979, Heinz Center.

9. Seifer, *Wizard*.

10. FBI's Nikola Tesla Memorandum, obtained via FOIA in Seifer, *Wizard*.

11. Kenneth Swezey notes, 1963, Heinz Center.

12. Anderson Cooper 360 Degrees, CNN, June 30, 2015.

13. Cheney, *Tesla: Man Out of Time*.

14. John Trump to Walter Gorsuch, January 30, 1943.

15. Seifer, *Wizard*, 454.

16. Exhibit F, Trump to Gorsuch, January 30, 1943.

17. Seifer, *Wizard*, p. 457.

18. Ibid., 460–61.

19. Leland Anderson to Kenneth Swezey, November 3, 1952. Swezey Collection, Smithsonian Institution.

20. http://www.fragmentsfromolympus.com/.

21. http://www.thebushconnection.com/bush.html.

22. Carlson, *Tesla: Inventor of the Electrical Age*, 404–05.

23. Tesla, *My Inventions*, 64.

24. Ibid., 35.

25. "Tesla, Man and Inventor," *New York Times*, March 31, 1895.

26. Tesla, *My Inventions*, 33.

27. "Radio Power Will Revolutionize the World," *Modern Mechanix & Inventions*, July 1934.

28. Tesla, *My Inventions*, 73.

29. Ibid., 77.

30. Leland Anderson, "Nikola Tesla—Last of the Pioneers?" *Journal of Engineering Education*, June 1959.

31. John Hays Hammond, Jr., to Kenneth Swezey, October 26, 1956.

32. Walter Stephenson, "Nikola Tesla and the Electric Light of the Future," *Scientific American Supplement*, March 30, 1895.

33. Seifer, *Wizard*.

34. Kenneth Swezey to Nikola Tesla, July 7, 1924.

35. O'Neill, *Prodigal Genius*.

36. Quote by John O'Neill from Cheney, *Tesla: Man Out of Time*.

37. Seifer, *Wizard*.

38. Brisbane, "Our Foremost Electrician."

39. Tesla, *My Inventions*, 63.

40. Graham Moore, *The Last Days of Night* (New York: Random House, 2016).

41. Cheney, *Tesla: Man Out of Time*, 328.

42. John Stone Stone in 1915, quote in Seifer, *Wizard*, 396.

43. "Tower to the People," a film by Joseph Sikorski.

44. Dragislav Pertkovich, "A Visit to Nikola Tesla," *Politika*, March 17, 1927.

45. Ibid.

46. Margaret Storm, *Return of the Dove* (Baltimore, MD: Margaret Storm, 1956).

47. Arthur Williams, *Wall of Light: Nikola Tesla and the Venusian Spaceship* as noted in Seifer, *Wizard*, 468.

48. Seifer, *Wizard*.

49. Robert Anton Wilson, *Cosmic Trigger: The Final Secret of the Illuminati* (New York: Pocket Books, 1975).

50. http://www.teslasociety.com/tunguska.htm.

51. Daniel Michaels, "Long Dead Inventor Nikola Tesla Is Electrifying Hip Techies," *Wall Street Journal*, January 14, 2010.

52. "Tower to the People," a film by Joseph Sikorski.

53. *Science*, May 16, 1958.

APPENDIX: THE MARVEL OF ELECTRICITY

1. Carl Van Doren, *Benjamin Franklin* (New York: Viking Press, 1938).
2. http://engineering.mit.edu/ask/what%E2%80%99s-difference-between -ac-and-dc.
3. Silverberg, *Light for the World.*
4. David E. Nye, *Electrifying America* (Cambridge, MA: The MIT Press, 2001).
5. Ibid.
6. Thomas J. Schlereth, *Victorian America: Transformations in Everyday Life* (New York: HarperCollins Publishers, 1991).
7. Christopher Cooper, *The Truth About Tesla* (New York: Race Point Publisher, 2005).
8. https://en.wikipedia.org/wiki/Joule.
9. Moore, *The Last Days of Night.*
10. Nikola Tesla, London Speech, 1892.

INDEX

ABOUT THE AUTHOR

Richard Munson is the author of several books, including *From Edison to Enron* and *Cousteau: The Captain and His World*. Based in Chicago, Illinois, he directs Environmental Defense Fund's clean-energy efforts in the Midwest. For more on the author, visit his website at www.richardmunson.com, Twitter @dickmunson, or LinkedIn at Dick Munson.